Wilhelm Döllen

The portable transit Instrument in the vertical of the pole Star

Wilhelm Döllen

The portable transit Instrument in the vertical of the pole Star

ISBN/EAN: 9783337025106

Printed in Europe, USA, Canada, Australia, Japan

Cover: Foto ©berggeist007 / pixelio.de

More available books at **www.hansebooks.com**

CINCINNATI OBSERVATORY,

Oct 13th 1870

The undersigned, on his own behalf ~~of this Observatory~~, begs leave to request the acceptance of the accompanying works.

Cleveland Abbe

Director of the Cincinnati Observatory.

ERRATA.

PAGE	LINE	FOR	READ
7	35	third	thread
8	16	than demanded	than is demanded
13	19	pure	mere
13	33	$\eta=o$	$\eta_0=o$
16	4	otherwhere	elsewhere
20	36	its principal step	in its principal steps
22	33, 34	thereby measured labor proves itself	increase of labor thereby produced proves
23	48	here	herewith
24	24	has	have
25	9	a (in 1st column)	a
27	—2	3640 (" " ")	3540
28	8	219 (" 7th ")	210
28	29	13720 (" 5th ")	12720
28	47	345 (" 7th ")	344
33	3	clock-chronometer	clock or chronometer
33	9	right	night
33	12	swerving	reversing
33	53	(Omit the last comma and close up the line.)	
34	34	near one thread	over one thread
35	37	for either of	for the first three of
36	25	methods	method
37	42	$b_1+b_0+\triangle_2p+\triangle_1p$	$b_1=b_0+\triangle_1p+\triangle_2p$
38	7	Erase and insert: $2c=(i_2-i_1)+(b_2-b_1)\cos Z-(T_2-T_1)\frac{dA}{dt}\sin Z$	
38	48, 49	$\sin a$ (and) $\cos a$	$\sin d$ (and) $\cos d$
39	11	J=&c.	t=&c.

PAGE	LINE	FOR	READ
39	45	quantity ,n	quantity n
39	—3	$(\cot g\,\phi - tg\,\delta)$	$(\cot g\,\phi - tg\,\delta\,\cot g^2\phi)$
40	19	$\nu = tg\,\phi\,\cot g\,\delta$	$\nu = tg\,\phi\,\cot g\,\delta^1$
40	47	2 $\cot g^2$	2 $\cot g^2\phi$
41	29, 39	$u_2 \pm C_2c$	$u_2 \mp C_2c$
42	12	factors	functions
42	14	Wrangel	Westphal

GENERAL STAR CATALOGUE.

STAR NO.	COLUMN	FOR	READ
4, 29, 53	3	Hydræ	Hydri
59	5	20^m	29^m
92	3	Geminorum	E Geminorum
109	3	ρ Cancri	6 Cancri
112	3	ι (15) Argus	15 (ι) Argus
154	5	27^s	25^s
170	3	a^1 Crucis	a Crucis
170	7	2′	22′
171	3	b Corvi	δ Corvi
199	3	η Bootis	θ Bootis
200	7	30″	36″
219	3	λ^2 Ursæ Minoris	γ^2 Ursæ Minoris
251	7	-31^0	$+31^0$

PORTABLE TRANSIT INSTRUMENT

IN THE

VERTICAL OF THE POLE STAR,

TRANSLATED

FROM THE ORIGINAL MEMOIR OF WM. DÖLLEN,

BY

CLEVELAND ABBE,

DIRECTOR OF THE CINCINNATI OBSERVATORY.

[with [illegible]

WASHINGTON.
GOVERNMENT PRINTING OFFICE.
1870.

NOTE.

The accompanying translation aud Appendix and Tables were prepared for the Bureau of Navigation in May, 1868. The proof-sheets have beeu kindly read by Professor Asaph Hall, of the Naval Observatory. A second memoir of Mr. Döllen upon this subject may be shortly expected aud will be at once translated and published.

CINCINNATI, *January*, 1870.

ON THE DETERMINATION OF THE TIME BY MEANS OF THE PORTABLE TRANSIT INSTRUMENT MOUNTED IN THE VERTICAL OF THE POLE STAR.

BY WILLIAM DÖLLEN,

SENIOR ASTRONOMER AT THE IMPERIAL CENTRAL OBSERVATORY, POULKOVA.

1. The method of determining the time, which is now to be considered, was proposed long since, and earnestly recommended by the most approved judges; it has been, indeed, developed in different works, sometimes with great thoroughness. Nevertheless, the fact remains that it has in nowise found that extensive dissemination and repeated application that it incontestably deserves.

The principal reason for this disregard is certainly to be found in this, that we have always regarded it as a kind of *dernier resort* to which we could at any rate have recourse, when perhaps circumstances would not allow of the generally more appropriate establishment in the meridian; moreover, the observations in and of themselves offered many inconveniences, and demanded a certain experience on the part of the observer in order to overcome difficulties which partly were inherent in the construction of our portable transits, partly, also, on the other hand, arose from the opinion that we must, for the sake of simplicity of computation, set for ourselves certain limitations. And moreover, especially when one desired the attainment of the finest possible results, the computations seemed so much more laborious than for the establishment in the meridian that the advantage of the shorter duration of the observations would be thereby more than counterbalanced.

It is the object of the following lines to oppose the above opinion and to give expression to the conviction, based upon many applications, that the method of observation under consideration has indeed not only even now a much greater importance than in general seems to be attributed to it, but that it may at once be unconditionally pronounced as the most preferable, after that certain changes, to be more exactly described hereafter, have been introduced in the instrument. Only this method develops for us the full value of the instrument; only this affords, under all circumstances and with the least delay, as accurate a result as can anyhow be attained with the means at hand.

It seems precisely now* of earnest importance to gain for the preceding truth the acknowledgment due to it, since an exceedingly important application of the portable transit to the determination of time is in such immediate prospect that a decision upon the method to be followed therein ought not to be longer delayed.

2. For the orientation of a transit instrument on our northern hemisphere *α* Ursæ Minoris, the Pole Star, offers important advantages over every other star, by reason of its brightness and nearness to the pole, as well as by reason of the accuracy with which its place in the heavens can be given for any moment of time. Therefore, in mounting a portable transit for the purpose of determining the time, even if the mounting

* This memoir was first published in 1863.

in and of itself be sufficiently trustworthy, we shall often do well, for the sake of observing Polaris, to renounce the other advantages that accompany an observation made as near as possible to the meridian.

This will, however, be absolutely necessary, if perhaps, as only too easily happens with the traveler, the mounting is by no means solid enough to sufficiently assure the invariability of the instrument during the entire interval necessary to a complete time determination in the meridian. In fact, under such circumstances no other method remains than to limit the duration of the observations to the shortest possible time, and to effect this, therefore, it is necessary to not wait until the proper stars, especially those situated near the pole and necessary for orientation, have reached the meridian, but to leave the meridian and to seek the most appropriate of them all, Polaris itself, wherever it may be in its diurnal circle. The dexterity of the observer will then show itself in the shortness of the interval which elapses after or even before the transit of Polaris over one of the threads of the reticule and the observation of the transit of the time star proper, (of course on all or at least as many threads as possible;) and we know from much experience that with a little practice this interval need be only a few minutes. The observation in the first position is completed by a careful determination of the inclination of the horizontal axis, or, as is very desirable, by two determinations inclosing the observations of the stars, and whose agreement bears testimony to the actual invariability of the instrument. It is, however, important immediately to execute a similar series of observations in the other position of the instrument, in order to free the result from the influence of the error of collimation and the difference in diameters of the pivots, without being under the necessity of taking these quantities from other sources.

It is easy, and will certainly also be intended in these observations, to observe Polaris both times upon the same thread, by preference the middle thread, and the succeeding computation will, in fact, be thereby somewhat simplified. One easily sees, however, that so far as concerns the true object of the observation, this is quite without serious importance, while in the execution of the observation a great relief may result from not being bound to any such condition. This stands in connection, however, with certain imperfections of the transit instruments in their present construction, to which the attention of the observer deserves to be especially called.

3. The instrument to which the following remarks are especially applicable is the Ertel portable transit instrument, although they certainly have also a more general importance. This instrument, in different forms, but all of nearly identical construction, has attained an extensive distribution, and may, indeed, through the descriptions given in words and drawings in different authorities, be considered as generally known. It differs chiefly from the formerly much-used Troughton's transit in the broken telescope and the possibility of the motion about the vertical axis. The former secures a marked facility in the observation, especially in the neighborhood of the zenith, and by reason of the shortening of the supports of the horizontal axis conduces much to give a greater rigidity to the entire instrument. The other change had certainly as immediate object to facilitate the exact adjustment in any azimuth, especially in the prime vertical, but the slow-motion screw serving thereto and the careful divisions of the horizontal circle, for reading which four verniers are provided, as well as the circumstance that by the pressure of a supporting spring the motion of the limb, in respect to the alidade, can be facilitated at will, give reason to suspect that the additional design was

entertained of rendering possible the exact measuring of horizontal angles.

But since any such intention is, through the absence of the assurance (or watch) telescope, only to a limited degree attainable, the greater mobility in azimuth therefore directly and very seriously endangers the excellence of the instrument as a transit. The clamp that ought to hold fast the two circles, with reference to each other, and thereby, therefore, the moveable upper, with reference to the immoveable lower portion, performs this service very imperfectly, not only because it operates only on one point, and that a point on the circumference, but also because of the slow motion that is combined with it. Very soon, therefore, it was that two simple clamping screws, distant 180° from each other, were applied, which are to be tightened as soon as the instrument is brought into the proper azimuth; and these are quite well adapted to greatly increase the security of the mounting.

On the other hand, not only is the exact adjustment in a known azimuth, as given by the readings of the circle, made almost impossible by reason of these clamping screws, since their greater or less tightness is accompanied by sensible flexures and corresponding derangements in the azimuth, but, furthermore, directly through these flexures is the fulfillment of the other condition, that the inclination of the horizontal axis shall be always the least possible, made much more difficult. These clamp screws can, moreover, if they are not applied to the proper points, give rise to a further danger, against which one must be forewarned. Evidently they should only be placed at the points where the supports of the horizontal axis are. In the earlier instruments they are, indeed, always found at these points, and only recently they have been placed 90° distant therefrom, probably for the sake of greater convenience in the manipulation; this is, however, precisely where they least of all realize their object; for if the previously-mentioned supporting spring be even very slightly compressed, then, notwithstanding the tightening of the misplaced clamp screws, the supports of the Y's will remain in a more or less unstable position, and thereby is generated the danger that the inclination of the axis may change when the level is set upon it.

We have persuaded ourselves, by direct trials, that the danger is not a fancied one, but that in the above way very sensible errors can arise. These become very apparent if on such an instrument with the spring compressed a series of levelings in alternate positions is made, as though for the determination of the difference of the pivot diameters. In case the necessary care in the reversions has been taken, we may certainly receive very accordant but entirely deceptive values; for, the influence of the unequal pivot diameters will have entirely disappeared, in comparison with that of the unequal weights of the two sides of the instrument. Therefore it is indispensable that in using such an instrument the spring should be perfectly slack, in order that before the tightening of the clamp screws the upper portion may rest entirely around with its whole weight, even though thereby the slow motion in one direction will, to a great extent, refuse to work. On no account, however, should the circumspect observer neglect to make a special investigation with the object of determining whether the application of the level itself does not still alter the inclination of the axis.

4. What precedes will quite suffice to show that some experience is necessary in order successfully to conduct the series of observations in that rapid succession which constitutes the importance of this method, as also that it affords a very important relief to be allowed to observe the Pole Star on any thread, but especially upon different threads, in the

two positions. This latter finds its full importance if it is thereby made possible to preserve the former azimuth after the reversion. Independent of the convenience which thus results to the observer, since he needs only once to clamp his instrument and adjust it for the inclination of the axis, the accuracy of the observation itself will thereby undoubtedly be increased, since it is well known that for some time after every change in the position of the instrument there remains the danger of a reactionary change; and finally one finds a very acceptable control for the entire operation in that for each position the same azimuth should result, although so far as the determination of time is concerned it is only necessary that the position of the instrument should not have changed during the observation in each position by itself.

This, however, it must be confessed, at least for our present instruments, assumes that the Pole Star has a definite motion in azimuth. Therefore, in the immediate neighborhood of its elongation this advantage must be renounced, and we are reduced to the necessity of adjusting anew the instrument in each position upon the almost motionless star. For, that Polaris becomes in general in this portion of its daily orbit less proper for our object, and therefore may, or even must, be replaced by another star in the neighborhood of the pole, perhaps by δ Ursæ Minoris, is, as it seems, a wide-spread yet manifest error. This has probably arisen, first, from the fact that the determination of the moment when the star bisects the thread will truly be then very inaccurate, and that, secondly, the reduction of such a transit from the thread to the great circle will always be large, and, under some circumstances, infinite.

Now, as concerns the first point, we shall, on consideration, recognize that therein not only no disadvantage exists, but there even results an advantage to the actual object of the observation, since precisely then the orientation of the instrument is achieved as accurately as the optical force of the telescope will at all allow; and in reference to the second point it is to be remarked, that this is a difficulty existing only in a certain method of computation, and therein is a proof that this method is not the proper one. No, if we are to leave the meridian at all it must be only for the sake of the advantage which the Pole Star offers; there is really no sufficient reason for choosing any other star for the orientation than Polaris.

At the same time, however, it is not to be denied that the observation in the neighborhood of the elongation has its special difficulties. We must, that is to say, not wait until the star, by its motion, is brought upon the thread, but must adjust the instrument directly upon the star, while, as above remarked, the azimuth will again be changed by the fastening of the clamping screw. With some practice, however, this difficulty is overcome, in that the clamp screw itself can, up to a certain point, replace the slow motion; but it is thereby always necessary carefully to watch that the simultaneously changing inclination of the horizontal axis remains small enough to allow of its accurate measurement. Furthermore, the correction of the inclination by means of the foot screws offers a means of changing a little the position of the star with reference to the threads, which can be occasionally very serviceable to the expert observer for the attainment of his object, especially if the construction of the instruments allows the level to remain upon the axis during the observation. Still another means of facilitating the observation at the elongation consists in the use of a somewhat inclined thread; we have, however, on this point no experience as to how advantageous this would prove itself in practice.

5. I allow myself, finally, two further remarks concerning the arrange-

ment of the observations, regard to which can be of practical importance under certain conditions. If we relinquish the maintenance of the same azimuth in both positions of the instrument, it is then occasionally possible to secure the other, in some circumstances not unimportant, advantage of being able to observe the same time star in both positions. This is especially important if no other time star of sufficient brilliancy is at hand; so that one must, without this resort, content one's self with observations in one position. The application of such a method fails on the approach of the time star to the zenith, on account of its increased azimuthal change. Neglecting the error of collimation and the advance of the Pole Star in azimuth, both which can affect the circumstances as well favorably as unfavorably, then in the latitude of 60° a change in azimuth of nearly 1° corresponds to a thread interval of 15′; that is to say, after the reversion the instrument must be changed in azimuth by this quantity in order still to be able to observe the Pole Star on the same thread; and it is easy to decide what change in the hour angle of the time stars corresponds to this change in azimuth. For α Bootis, for instance, it amounts to nearly three minutes of time, and it depends only upon the expertness of the observer and the arrangement of the instrument whether this interval suffices for the reversion and adjustment in azimuth.

The other remark refers to the difficulty known to all observers of observing the transit of Polaris if the star is very faint. Often one catches the star very fairly at some distance from the thread, while on approaching it it completely disappears. The observation is then usually made by noting the disappearance of the star on one side of the thread and its reappearance on the other, and the mean of both moments is considered as the time of transit. Of course, however, such an observation is far less accurate than if the star remains visible upon the thread itself, and it will therefore be the less available the longer the interval between the disappearance and reappearance. Now, we have in such cases, with the best results, replaced the observations upon the thread by the observation in the middle between two threads, and find that with an appropriate ~~third~~ interval there results thereby scarcely an appreciably diminished accuracy of observation.

6. If we review the previous remarks it results that for the determination of time by means of a portable transit instrument, the method of observation here considered is only in very special cases inferior to the observation in the meridian; on the other hand, in other circumstances which far more frequently present themselves, it will, by this method only, be possible to realize any determination at all; and therefore the advice seems to be authorized, that we do not shun the labor of putting ourselves by means of the necessary practice, to a certain extent at least, in possession of all the means which the instrument can offer to the observer to whom it is intrusted.

The judgment must, however, be quite different if it should be possible to give to the instrument an arrangement by which the various difficulties, above mentioned, are still further removed from the observation. And this is now in reality very completely attainable, and that through changes which have already proved themselves as individually quite applicable, although their aggregate combination has still first to stand the proof of actual experience. Now, for our purpose, the most important addition to our present transit instruments seems to me to be the introduction of a movable thread, whose position, with respect to the fixed threads, will be any time recognized by means of the micrometer screw that carries it. This change, since it only affects the ocular, is also comparatively easy to effect in the existing instruments. The decisive success which is actually attained with the large fixed instruments in different places shows that by proper care and circumspection in the application and use of such a micrometer, an accuracy and reliability can be attained that quite suffices for our purpose. But the importance due to the movable thread in the portable instrument will be only very imperfectly estimated by the advantage that has resulted from its application to the large fixed instruments; for, important as may be, for these latter instruments, the facility of observation thereby secured, the method of observing remains unchanged, since, if we will free ourselves as completely as possible from the assumption of the invariability of the mounting, then a meridian mark is indispensable. On the other hand, the movable thread, to a certain degree, provides a meridian mark for the portable instrument. This is the Polar Star itself, that can now on any point of its daily orbit be with equal convenience and reliance observed without a greater consumption of time than demanded to direct the telescope and point the micrometer thread upon the star, precisely as if a mark were observed.

Together, however, with this important auxiliary to the achievement of an accurate orientation of the instrument at any instant, provided always the optical force of the instrument allows the Pole Star to be actually perceived, are still two further changes in the instrument extremely desirable, if we would impart to the entire determination of time the greatest degree of security and convenience. That is to say, it is to this end necessary, first, that not only during the observation itself but also during the reversion of the instrument the level should remain upon the axis; and second, that this reversion should be accomplished much more rapidly, and especially more safely—that is to say, without danger of any other change than is attainable in the reversion with the unaided hand.

The possibility of a perfect execution of these arrangements, even in smaller instruments, is circumstantially shown by experience; but truly they affect the construction of the entire instrument so vitally that they must, in planning it, necessarily be kept in view. The increased labor and care to be expended in the construction of such an instrument, and the increased costliness dependent thereon, should not, it seems, come into consideration in contrast with the important increase in the value of the observations resulting therefrom. Such expenditure will be, even in a brief use, richly repaid by the saving in labor at every application, without its being necessary especially to mention the further advantage that is found in the diminution of the danger of seeing the series of observations that has been commenced interrupted before its completion.

How important in reality is the diminution of the duration of the observations which results from our method with instruments specially constructed therefor, will be recognized if one once considers the course pursued in a complete determination of time. As to what concerns the preparation for the observation, that consists mostly in clamping the instrument with sufficiently small inclination of the axis in such an azimuth that the Pole Star appears approximately in the middle of the reticule. As soon, then, as the observation of the transit of a proper time star is finished, the level and Pole Star are read off; a reversion of the level upon the axis is not necessary if we at once reverse the instrument itself and thereby intentionally disturb neither inclination nor azimuth. A second time star in this other position of the instrument with the thereto belonging readings of level and Pole Star, completes the determination of time.

It is seen that the duration of the whole depends mostly only upon how quickly the proper time stars follow each other in the heavens, and will, therefore, in general, not need to exceed a very small number of minutes. Finally, however, it is, after all, not this decided saving of labor, important as it in itself may be, that gives the desired contraction of the duration of the observations its proper value, but this rests rather upon the increase of the accuracy that arises therefrom. I estimate this to be very considerable. But to enable any such opinion to claim a universal recognition, it must be based upon unalterable numerical values, which can only be obtained by actual application continued for a long time and under the greatest variety of exterior circumstances. Such numbers are at this moment not at my command, and I therefore content myself with illustrating, by the following considerations, the opinion that I have formed after many years' experience in the matter.

One's first thought may be to increase the accuracy of the time determination, if not to any at least to a considerably higher degree, by the repetition of the observations. If we estimate about fifteen minutes for the duration of a complete determination, which, according to what precedes, is considerably more than what will generally be necessary, then, in an hour, which in general certainly would not suffice for a complete time determination in the meridian, this may be four times repeated, whereby the probable error of the final clock correction, in so far as it depends upon the astronomical observation, would be reduced to one-half. Now I am of the opinion that this would be practically of no use whatever. That is to say, I think that already the single determination possesses such a degree of accuracy that, independent of error in the star places, it cannot be further increased by multiplication of observations; even the circumstance that a uniformity in the chronometer rate must then be assumed for so much longer an interval, suffices alone to annul the advantage of reducing by one-half the error of observation. I mean, in all seriousness, that the probable error of such a time determination, so far as the observation has an influence on it, depends principally upon nought else than the accuracy of the observed transits over the single threads, which evidently is the limit of accuracy that any method whatever can afford. And should this, as I confidently hope, be confirmed by unprejudiced examination, then will the expression not seem too venturesome that in the construction of a portable transit instrument, of moderate dimensions, according to the principles above laid down, and in artistic perfection, such as we shall in a short time receive from the hands of Brauer, a new epoch begins in the solution of that very important problem, the absolute determination of time.

7. Finally, I cannot refrain from noticing with a few words more particularly the especial occasion which has been before alluded to, and which renders it extremely desirable that precisely now the great advantages of the method of determining time here considered should find their proper acknowledgment and be made useful in the practical application. This is the measure, now approaching its completion, of a parallel of the earth under the 52° of north latitude, embracing an arc of 68° 54′, that truly magnificent international undertaking, to have given a start to which is a fit conclusion of the labors by which W. Struve has so ably promoted geodesy during his whole life.

Already is the determination of the astronomical differences of longitudes along the entire arc appointed for the next two years. The transmission of time will everywhere, when possible, be made by means of the electric telegraph. Since, however, at present the telegraph does not directly connect the different points of the arc, the desired differences of longitude will be obtained by means of the central telegraph stations of the different States. Thus it will be possible that all the time determinations upon the entire arc shall be made by the same observer with the same instrument; and furthermore, it is provided that this shall be entirely independently carried out by the two different observers, so that for each individual difference of longitude the necessary clock correction shall depend upon the quite independent time determination of each observer. It is, however, equally designed that each observer should be assigned to one and the same instrument. Now, this instrument will indeed, in order to facilitate the labor, be provided with a level that does not need to be removed during the observation, and with a special apparatus for reversion. And still I hold decidedly to the opinion that, independent of other difficulties, the fulfillment of the programme for any night's observation, under the conditions connected with a complete time determination in the meridian, will be possible only under very specially favorable, and therefore seldom occurring, exterior conditions, especially if we, as certainly must be considered necessary, will not relinquish the condition that the transmission of time be each time, and as closely as possible, included by time determinations at each place.

Now, under these circumstances, the method here recommended seems to be an especially fortunate expedient. Assuming our reticule to consist of seven threads, then the two stars twice observed by each observer give a sum total of fifty-six individual thread transits, and thus a time determination that is quite comparable with a time transmission by means of as many telegraphic signals. And the whole time necessary for this operation need be scarcely an hour. It seems useless to attempt to increase the accuracy of the determination for one night by repetition, as is easily possible; imperative, however, is the repetition on different nights, as is also prescribed in the adopted programme. The final determination of the details in this respect, as also in general, properly remains deferred until the accurate investigations which will be made in the immediate future, as preparatory to the work of the next years, be completed. For the present we are concerned only with the acknowledgment of a fundamental principle which I would thus enounce:

FOR THE ATTAINMENT OF A TIME DETERMINATION AT A GIVEN MOMENT, THE PORTABLE TRANSIT INSTRUMENT WILL, UNDER ALL CIRCUMSTANCES, BE BEST MOUNTED, NOT IN THE MERIDIAN BUT IN THE VERTICAL OF THE POLE STAR.

The next months will bring us the final practical proof of the truth of this thesis.

8. When the question as to the practical value of the method of observation here considered is once settled, then will the overcoming of the greater labor of computation be an object of comparatively secondary importance. A nearer consideration shows, however, that this is by no means so excessive as it appears at the first view; and the relation is even reversed in those cases where, for the meridian time determination, we dare not be contented without the exact solution according to the method of least squares.

The deduction of methods of computation which shall unite the greatest convenience of execution with the necessary accuracy of the results, has been undertaken at different times and in different places, especially lately on the part of Hansen, in an exhaustive manner, in an article in No. 1136, Vol. XLVIII, of the Astronomische Nachrichten. Starting from the known general equation for the transit instrument, Hansen gives first a direct exact solution by purely analytical means, which is well worthy of consideration by reason of the art with which the unknowns are found from the transcendental equations. Since, however, there can never be sufficient ground to make use of this exact but, therefore, more tedious solution, there are next deduced approximate formulæ for actual use, which really leave scarcely anything more to be desired. Although after this every further discussion of the subject might seem idle, yet I venture in what follows to present the somewhat different method that I for years have followed in treating this problem, and that in very frequent applications has always proved itself to my mind thoroughly appropriate. The direct exact solution to which I arrive at first seems to me to recommend itself as very advantageous, by reason of the exceeding clearness of the geometrical considerations on which it is based; afterwards, however, this leads to approximate formulæ and precepts for computation which are so simple and convenient, and especially so entirely free from uncertainty in reference to the signs of all the quantities coming under consideration, that they perhaps deserve consideration even when compared with Hansen's.

9. For greater generality, and in conformity with the above remarks, I in the following development assume that the two stars are not observed on the same thread. After what more immediately follows will be found the reduction that is necessary if the time star is observed on more than one thread. For the present the thread on which the time star is observed is called the middle thread, and the thread for the Pole Star is distant by f from this; thus, $90° + c$ and $90° + c + f$ designate the distance from the west end of the horizontal axis of the transit instrument of the celestial points determined by these two visual lines.

The exact solution of our problem consists fundamentally in this, to find the diagonal WP and the angle WPS of a spherical quadrilateral WSPS′, in which the four sides and the angle P are given. Let us represent by W and P the points in the heavens to which the axes of revolution of the instrument and of the earth are directed, and, to leave nothing uncertain, the west pole of the instrument and the north pole of the equator; S and S′, on the other hand, the places of the Time Star and of the Pole Star at the moment of observation; we thus know at first,

$$WS = 90° + c, \qquad WS' = 90° + c + f,$$
$$PS = 90° - \delta, \qquad PS' = 90° - \delta'.$$

Finally, the angle SPS′, which we will designate by τ, would be simply the difference of the right ascensions of the two stars, if both were observed at the same moment. Since, now, some time at least must elapse between these two observations, this difference is still to be changed by the hour angle corresponding to this interval; and in this nothing further is assumed than that we know the clock rate sufficiently well to make this change in hour angle with an accuracy corresponding to the accuracy of the observations. If, however, the observations are arranged as is proper, then even without the movable thread would this interval never be greater than from five to six minutes, and then a correction on account of the clock rate would scarcely be further demanded, even if the chronometer kept mean time instead of sidereal time; the estimation of the moment when the Pole Star is on the thread is assuredly, with a portable transit instrument, not certain to within a second, even in the neighborhood of the culmination. The concluded clock correction belongs, of course, to the moment of observation of the time star.

Let a great circle be drawn through the points S′ and S; then is the exact deduction of W P and the angle W P S given by the following equations:

$$\text{(I)}\ \triangle\, SPS':\begin{cases}(1)\ \sin SS' \sin PSS' = \sin PS' \sin SPS' \\ (2)\ \sin SS' \cos PSS' = \sin PS \cos PS' - \cos PS \sin PS' \cos SPS' \\ (3)\ \cos SS' = \cos PS \cos PS' + \sin PS \sin PS' \cos SPS'\end{cases}$$

$$\text{(II)}\ \triangle\, WSS': \qquad \cos WSS' = \frac{\cos WS' - \cos WS \cos SS'}{\sin WS \sin SS'}$$

$$\text{(III)}\ \triangle\, WPS:\begin{cases}(1)\ \sin WP \sin WPS = \sin WS \sin WSP \\ (2)\ \sin WP \cos WPS = \sin PS \cos WS - \cos PS \sin WS \cos WSP \\ (3)\ \cos WP = \cos PS \cos WS + \sin PS \sin WS \cos WSP\end{cases}$$

We will now introduce the above-chosen notation, δ, δ', c, f, τ, into these equations, and also the following:

$$SS' = 90^\circ - d, \qquad PSS' = \xi;$$
$$WSS' = 90^\circ + \eta;$$
$$WP = 90^\circ + n, \qquad WPS = 90^\circ - x;$$

whose immediate object is merely to indicate at once such quantities as, in the application, can never be other than small. In case the arc S S′, or the distance of the two points in which the stars are observed, is neither very near 180° nor near 0°, both which cases are excluded by the nature of the problem, then is η of the same order as c and f, while ξ, n, x, as also the difference $d - \delta$, are of the same order as P S′.

But, entirely without reference to the magnitudes of all quantities, our exact formulæ are now the following:

$$\text{(I)}\ \triangle\, SPS':\begin{cases}(1)\ \cos d \sin \xi = \cos \delta' \sin \tau \\ (2)\ \cos d \cos \xi = \cos \delta \sin \delta' - \sin \delta \cos \delta' \cos \tau \\ (3)\ \sin d = \sin \delta \sin \delta' + \cos \delta \cos \delta' \cos \tau\end{cases}$$

$$\text{(II)}\ \triangle\, WSS': \qquad \sin \eta = \frac{\sin (c + f) - \sin c \sin d}{\cos c \cos d}$$

$$\text{(III)}\ \triangle\, WPS:\begin{cases}(1)\ \cos n \cos x = \cos c \cos (\xi + \eta) \\ (2)\ \cos n \sin x = - \cos \delta \sin c + \sin \delta \cos c \sin (\xi + \eta) \\ (3)\ \sin n = \sin \delta \sin c + \cos \delta \cos c \sin (\xi + \eta)\end{cases}$$

Thus far, as is seen, everything is entirely independent of any reference to the place of observation upon the surface of the earth, which reference will be first obtained by the knowledge of the position of the zenith, Z, with respect to the quadrangle hitherto considered. To this end are used the distances of the point Z from the points P and W, which are given by the latitude of the place and inclination of the horizontal axis of the instrument. If we put, as usual,

$$ZP = 90^\circ - \varphi, \qquad ZW = 90^\circ - b,$$

where b is positive if the west end of the axis lies above the horizon; also, further,

$$ZPW = 90^\circ - m, \qquad PZW = 90^\circ + a,$$

there results:

$$\text{(IV)}\ \triangle\, WPZ:\begin{cases}(1)\ \sin m = \operatorname{tg} n \operatorname{tg} \varphi + \sin b \sec n \sec \varphi \\ (2)\ \sin a = \operatorname{tg} b \operatorname{tg} \varphi + \sin n \sec b \sec \varphi\end{cases}$$

by means of which our problem is now completely solved. For

$$(90^\circ - m) - (90^\circ - x) = x - m = 15\,t$$

is the western hour angle of the time star at its observation at S, and consequently the sidereal time of this observation is

$$a + t = S + u,$$

if S signifies the observed clock time and u the correction at this moment of the chronometer to sidereal time. Therefore, finally,

$$u = \left(\frac{x - m}{15}\right) - (S - a).$$

10. Should one at any time have reason to make use of these exact formulæ, I would recommend that they be applied as they here stand, without thinking of the introduction of auxiliary angles. These auxiliary angles have, for the computations of the present day—thanks to the increasing dissemination of the Gaussian logarithms—lost, to a great extent, their former importance; they afford a real relief in the computation generally, only when we have to do, not with a single case but with many connected together, to which certain quantities are common, as, for example, often in the computation of tables.

On the other hand, the experienced computer will certainly not fail to make full use of the advantages offered by the smallness of most of the desired quantities, and that, too, without impairing in the least the perfect exactness of the result.

Here, however, such a computation has always a ~~pure~~ mere theoretical interest. In actual practice approximate formulæ suffice, as above several times remarked, even in the extremest cases which can be likely to occur. These approximate formulæ are obtained, if we derive the desired quantities from the above equations, under the assumption that $b = 0$, $c = 0$, $f = 0$, and thereupon develop the coefficients of these quantities, with whose assistance we can at pleasure compute their influence if they are not equal to zero, a method of solution which will be necessary, in reference to c, as soon as the error of collimation is not assumed as known, but must first be found from these same observations themselves.

If we indicate by the subscript o the values resulting under the assumption mentioned, and by a prefixed Δ the correction then necessary, so that, for instance, $u = u_0 + \Delta u$, the above given strict formulæ give, without trouble, the following:

$$\eta_0 = 0 \qquad\qquad \Delta\eta = \frac{1 - \sin d}{\cos d} \cdot c + \frac{1}{\cos d} \cdot f$$

$$\begin{aligned}
\cos n_0 \cos x_0 &= \cos\xi \\
\cos n_0 \sin x_0 &= \sin\delta \sin\xi & \cos n \cos x \,.\, \Delta x - \sin n \sin x \,.\, \Delta n &= -\cos\delta \,.\, c + \sin\delta\cos\xi \,.\, \Delta\eta \\
\sin n_0 &= \cos\delta\sin\xi & \cos n \,.\, \Delta n &= \sin\delta \,.\, c + \cos\delta\cos\xi \,.\, \Delta\eta
\end{aligned}$$

$$\begin{aligned}
\sin m_0 &= \operatorname{tg}\phi \operatorname{tg} n_0 & \cos m \,.\, \Delta m &= \sec n \sec\phi \,.\, b + \sec^2 n \operatorname{tg}\phi \,.\, \Delta n \\
\sin a_0 &= \sec\phi \sin n_0 & \cos a \,.\, \Delta a &= \operatorname{tg}\phi \,.\, b + \cos n \sec\phi \,.\, \Delta n
\end{aligned}$$

Hence, with help of equations I, there follows:

$$\operatorname{tg} x_0 = \sin\delta \operatorname{tg}\xi = \frac{\sin\delta\cos\delta'\sin\tau}{\cos\delta\sin\delta' - \sin\delta\cos\delta'\cos\tau};$$

$$\operatorname{tg} n_0 = \operatorname{cotg}\delta \sin x_0;$$

$$\sin m_0 = \operatorname{tg}\varphi \operatorname{cotg}\delta \sin x_0.$$

If, therefore, we put

$$\operatorname{tg}\delta \operatorname{cotg}\delta' = \lambda, \quad \operatorname{tg}\varphi \operatorname{cotg}\delta = \mu;$$

then will

$$\operatorname{tg} x_0 = \frac{\lambda \sin\tau}{1 - \lambda\cos\tau}, \quad \sin m_0 = \mu \sin x_0;$$

and finally,

$$u_0 = \left(\frac{x_0 - m_0}{15}\right) - (S - a).$$

But,

$$u = u_0 + \frac{\Delta x - \Delta m}{15};$$

therefore,

$$u_0 = u + \frac{\Delta m - \Delta x}{15};$$

for which we will write,

$$u_0 = u + \mathrm{B}b + \mathrm{C}c + \mathrm{F}f,$$

in which the divisor 15 in the coefficients B, C, F, which are now to be immediately developed, need not be further considered if we have expressed b, c, f, in time. We remark, first, in reference to this development, that in the above-given differential formulæ the cosines of the small angles, b, c, f, have been put equal to unity, and the sines proportional to the angles themselves, while in the case of the small quantities, ξ, x, n, m, a, this simplification has not yet been introduced. This is proper, since that these latter, although small in themselves, are, or at least can become, many times larger than the former. Of course b will naturally amount only to a few seconds of arc, since larger inclinations cannot be measured with the level. The intelligent observer will furthermore take care that c be always very small, but it is not at all meant to advise to attempt to make it zero every time. Finally, in the most extreme cases f can certainly be so large as 15 minutes of arc; the extreme threads will certainly never stand further than this from the middle one; but even then cosine f does not differ from Radius by a unit in the fifth decimal place. We will most easily receive an idea of the magnitudes of the other quantities if we represent their approximate values as functions of any one, most conveniently of n. Our formulæ give us at once:

$$a = n \sec\varphi, \quad m = n \operatorname{tg}\varphi, \quad \xi = n \sec\delta, \quad x = n \operatorname{tg}\delta.$$

Therefore, if we put

$$1 - \cos n = p,$$

there results, approximately,

$$1-\cos a = p\sec^2\varphi, \; 1-\cos m = p\operatorname{tg}^2\varphi, \; 1-\cos\xi = p\sec^2\delta, \; 1-\cos x = p\operatorname{tg}^2\delta;$$

further,

$$\sin n = \sqrt{2p}, \quad \sin x = \operatorname{tg}\delta\sqrt{2p},$$

therefore,

$$\sin n \sin x = 2p \operatorname{tg}\delta;$$

and we will now easily be able to judge how far these small fractions still demand notice. Since we adopt the rule to use no other star than the Pole Star for orientation, therefore n can never be greater than 1° 26′, to which corresponds $p = 0.0003$; and therefore for a latitude of 60° and the observation of zenith stars, none of the above quantities become greater than about 0.001. Now this is entirely evanescent, both in respect to b and to c, but not in respect to the extreme value of f, especially in case we would, as we ought, retain the hundredths of a second of time in our results. If we therefore may neglect these fractions at first, for the sake of perspicuity, in the deduction of the coefficients, B, C, F, still we must finally develop for F the small corrections demanded by this omission in order that this may always, when necessary, be brought into account. The differential equations, simplified by the first mentioned assumption, now are:

$$\Delta x = -\cos\delta \,.\, c + \sin\delta \,.\, \Delta\eta$$
$$\Delta n = +\sin\delta \,.\, c + \cos\delta \,.\, \Delta\eta$$
$$\Delta m = +\sec\varphi \,.\, b + \operatorname{tg}\varphi \,.\, \Delta n.$$

From these, by substitution of $\varDelta n$ from the previous equations, there results,

$$\varDelta m = \sec\varphi \,.\, b + \operatorname{tg}\varphi \sin\delta \,.\, c + \operatorname{tg}\varphi \cos\delta \,.\, \varDelta\eta\,;$$

therefore,

$$\varDelta m - \varDelta x = \sec\varphi \,.\, b + (\operatorname{tg}\varphi \sin\delta + \cos\delta)\,.\, c + (\operatorname{tg}\varphi\cos\delta - \sin\delta)\,.\,\varDelta\eta,$$

or

$$\varDelta m - \varDelta x = \sec\varphi \,.\, b + \sec\varphi\cos z \,.\, c + \sec\varphi \sin z \,.\, \varDelta\eta,$$

where we have put

$$\varphi - \delta = z.$$

Finally, if we substitute the value

$$\varDelta\eta = (\sec d - \operatorname{tg} d)\,.\, c + \sec d \,.\, f,$$

and unite the two terms dependent upon c, the desired coefficients result,

$$\begin{aligned} B &= \sec\varphi \\ C &= \sec\varphi \sec d\,[\sin z + \cos(d+z)] \\ F &= \sec\varphi \sec d \sin z + Gp, \end{aligned}$$

where G is a factor still to be deduced, namely, the coefficient of the hitherto neglected term in the value of F dependent upon p. In order to obtain this we must return to the complete differential equations and seek the value of the partial differential coefficients with reference to f, which we will designate by $d\,x$, $d\,n$, $d\,m$; this will at the same time be a control over the accuracy of the previously-found principal terms of the factor F, independent of p.

We have

$$d\,\eta = \sec d,$$

and therefore

$$\begin{aligned} \cos n \cos x \,.\, d\,x - \sin n \sin x\, d\,n &= \cos\xi \sec d \sin\delta \\ \cos n \; d\,n &= \cos\xi \sec d \cos\delta \\ \cos m\, d\,m &= \sec^2 n \operatorname{tg}\varphi \,.\, d\,n. \end{aligned}$$

If, by means of the second equation, we eliminate the $d\,n$ from the two others, we obtain:

$$\begin{aligned} \cos n \cos x \,.\, d\,x &= \cos\xi \sec d \sin\delta + \operatorname{tg} n \sin x \cos\xi \sec d \cos\delta \\ \cos m \,.\, d\,m &= \sec^3 n \operatorname{tg}\varphi \cos\xi \sec d \cos\delta, \end{aligned}$$

or remembering that

$$\cos n \cos x = \cos\xi,$$

we have

$$\begin{aligned} d\,x &= \sec d \sin\delta + \operatorname{tg} n \sin x \sec d \cos\delta \\ d\,m &= \operatorname{tg}\varphi \sec^2 n \sec m \cos x \sec d \cos\delta, \end{aligned}$$

and therefore

$$d\,m - d\,x = F = \sec d \cos\delta\,(\operatorname{tg}\varphi \sec^2 n \sec m \cos x - \operatorname{tg} n \sin x - \operatorname{tg}\delta).$$

But, from the previous remarks, there follows:

$$\sec^2 n \sec m \cos x = 1 + p\,(2 + \operatorname{tg}^2\varphi - \operatorname{tg}^2\delta) + \ldots . \text{ \&c.}$$

$$\operatorname{tg} n \sin x = \frac{2\,p \operatorname{tg}\delta}{1-p} = 2\,p \operatorname{tg}\delta + \ldots\ldots \text{\&c.},$$

therefore,

$$\begin{aligned} F &= \sec d \cos\delta\,\{\operatorname{tg}\varphi - \operatorname{tg}\delta + p\,[2\operatorname{tg}\varphi + (\operatorname{tg}^2\varphi - \operatorname{tg}^2\delta)\operatorname{tg} - 2\operatorname{tg}\delta]\} \\ &= \sec d \cos\delta\,(\operatorname{tg}\varphi - \operatorname{tg}\delta)\,\{1 + p\,[2 + (\operatorname{tg}\varphi + \operatorname{tg}\delta)\operatorname{tg}\varphi]\} \\ &= \sec\varphi \sec d \sin z\,\{1 + p\,[2 + (\operatorname{tg}\varphi + \operatorname{tg}\delta)\operatorname{tg}\varphi]\}. \end{aligned}$$

The portion independent of p agrees, as it should, with that previously deduced. If we indicate it temporarily by F_0 and by h, the factor of p, then will

$$F = F_0(1 + ph) = F_0(1 + p)^h = F_0(\sec n)^h.$$

We now remark that sec n does not otherwhere occur in our computation, and that, therefore, it would be convenient to replace it in some way by sec m, which is found, without further trouble, in taking from the tables the angle belonging to m. From $m = n \operatorname{tg} \varphi$ there results

$$\sec n = (\sec m)^{\operatorname{cotg}^2 \phi}, \text{ therefore } (\sec n)^h = (\sec m)^{h \operatorname{cotg}^2 \phi},$$

thus, finally,

$$F = \sec \varphi \sec d \sin z (\sec m)^k$$

where

$$k = 1 + 2 \operatorname{cotg}^2 \varphi + \operatorname{cotg} \varphi \operatorname{tg} \delta.$$

A similar remark is to be made with reference to the arc d, the complement of the distance between the two points of the heavens S′ and S, in which the two stars are found at the time of their observation. This arc can be found, with all desirable accuracy, from the equations (I.) But in all cases actually occurring this arc passes, within a few seconds, through the zenith; and further, since it is only used in the computation of the differential coefficients C and F, it is more than sufficiently accurate if we put

$$90° - d = z' + z, \text{ therefore } d + z = 90° - z',$$

and take for z' the actual zenith distance of the Pole Star, counting it always positive, while z remains $= \varphi - \delta$. If, as is for these observations always desirable, we have an ephemeris of the Pole Star, that is to say, a table of azimuths and zenith distances with the argument, sidereal time, which indeed is indispensable if the observations are made by day or twilight, then z' can, at any time, be taken therefrom. If we have no such table, or none sufficiently accurate, we need only to read the vertical setting circle simultaneously with the observation; and if this be also done for the time star then the difference of the two readings, even without any knowledge of the position of the zenith, gives directly the complement of the arc d.

I will not here omit to call attention to a general remark that has very often proved itself to be one of great importance. We make it an invariable rule in the accurate observation of either co-ordinate also to approximately determine the other, which is always possible by means of the finding circle. Thus, in the observations of transits one always reads also the altitude circle; in measuring the zenith distances one reads also the horizontal circle; or rather, we do not neglect to record these readings, which will generally be made in order to set the instrument. Independent of the thus-secured sure solutions of many doubts arising in consequence of mistakes, &c., these quantities are always of importance, and have a direct application as soon as we have to do with the differential coefficients; and, although these latter are not always used in the deductions of the results themselves, it is still very desirable not to neglect their development, when not entirely too tedious, since they conduce very much to the attainment of a correct judgment as to the reliability of the determination obtained.

11. The following group gives a complete review of the process of the computation, according to the approximate formulæ just developed. Let there be—

For the Pole Star.	For the Time Star.	
S′	S	the clock time of observation.
$u + \gamma$	u	the correction to sidereal time.

For the Pole Star.	For the Time Star.	
α'	α	the apparent right ascension.
δ'	δ	the apparent declination.
z'	z	the zenith distance, z' is always positive, $z = \varphi - \delta$.

We now form the quantities:

$$S' + \gamma - \alpha' = D' \qquad \operatorname{tg}\delta \operatorname{cotg}\delta' = \lambda \qquad \operatorname{tg} x_0 = \frac{\lambda \sin \tau}{1 - \lambda \cos \tau}$$

$$S \qquad - \alpha = D$$

$$15\,(D' - D) = \tau \qquad \operatorname{tg}\varphi \operatorname{cotg}\delta = \mu \qquad \sin m_0 = \mu \sin x_0$$

thus we have:

$$u + B\,b + C\,c + F\,f = \frac{x_0 - m_0}{15} - D;$$

where

$B = \sec \varphi$ — b the inclination of the horizontal azis.

$C = \sec \varphi \dfrac{\sin z' + \sin z}{\sin (z' + z)}$ — c the error of collimation of the middle thread.

$F = \sec \varphi \dfrac{\sin z}{\sin (z' + z)} (\sec m)^k$ — f the distance of the middle thread from that on which the Pole Star is observed.

$$k = 1 + 2 \operatorname{cotg}^2 \varphi + \frac{1}{\mu}$$

The inclination b is positive if the west end of the axis lies above the horizon; the signs of c and f are to be understood so that $90^\circ + c$ and $90^\circ + c + f$ represent the distances of the respective points of the heavens from the west ends of the axis. All three, b, c, f, are expressed in time, and u will be obtained in time.

The introduction of the term Ff could evidently be avoided by always considering the thread on which the Pole Star is observed as the middle thread, and reducing the transit of the time star to this. Of course, then, the error of collimation would, in general, be variable both in sign and also in magnitude in the different positions by the known value of the distance of the middle threads adopted in the two positions. No advantage results, however, in practice from this consideration, correct as it is in and of itself. For, since in this case the collimation error could have a much larger value than otherwise need be assumed, therefore the exact development which is now considered necessary for the factor F would be demanded for the factor C, and this would lead to expressions still more complicated than those found for F. At least I have not myself been able to represent the correction demanded in such a case for the above-given value of c more conveniently than in the form of a factor,

$$(\sec m)^q, \text{ where } q = \operatorname{cosec}^2 \varphi + \operatorname{cotg}\varphi \operatorname{tg}\frac{z' - z}{2}\left(1 - \frac{1}{\mu}\right),$$

which factor is, therefore, to be made use of if ever the collimation error should have an especially large value.

If, moreover, many time determinations are to be made under the same latitude according to our method, it is then certainly most convenient to compute the factors C and F once for all, for the always moderate number of stars that can come into consideration; as we are, indeed, already accustomed to do for the much simpler factors that are used to free the transits observed in the immediate neighborhood of the meridian from the influence of the different instrumental errors.

12. The endeavor to further simplify the deduction of the clock correction by any application of serial developments has but little prospect of success; for, because of the magnitude to which the angles x and m can attain, so many members of the series are required in order to obtain accurate results, that the computation according to the exact formula is decidedly more convenient. If, however, we will be content with an accuracy of about $0^s.1$, we may write:

$$\log\left(\frac{m_0 - x_0}{15}\right) = \log(\beta \sin\tau) + \gamma \cos\tau,$$

where

$$\beta = \frac{\lambda(\mu - 1)}{15 \sin 1''} = \frac{\cot \delta' (\text{tg}\,\varphi - \text{tg}\,\delta)}{15 \sin 1''},$$

and

$$\gamma = \text{mod}\,.\,\lambda = 0.4343 \,\text{cotg}\,\delta' \,\text{tg}\,\delta.$$

Although these formulæ may be only seldom used in the proper computation of the observations, they at least offer the advantage of an easy insight into the dependence of the desired clock corrections upon the imperfections of the observations and the errors of the adopted elements of reduction.

If we also neglect the second term of the formula just given—as it evidently, for the present purpose, does not come into consideration—and, besides, allow ourselves to put

$$z' = 90° - \varphi, \text{ and therefore } d = \delta,$$

whereby in the worst case only so much will be lost as depends upon the polar distance of the Pole Star, we receive

$$\begin{aligned} -u = D &+ \frac{\text{cotg}\,\delta' (\text{tg}\,\varphi - \text{tg}\,\delta)}{15 \sin 1''} \,.\, \sin\tau \\ &+ b \,.\, \sec\varphi \\ &+ c \,.\, (\text{tg}\,\varphi + \sec\delta - \text{tg}\,\delta) \\ &+ f \,.\, (\text{tg}\,\varphi - \text{tg}\,\delta), \end{aligned}$$

which expression seems quite well suited to serve as a starting point for such considerations. We will, however, here content ourselves with noticing only that which has reference to the choice of the star which we thus far have spoken of under the name of the Time Star, without thereby having intended to determine anything further as to its place in the heavens. The first thought is to direct the choice by preference to equatorial stars in consideration of the, in itself, correct geometrical principle, that two points on the sphere determine the corresponding great circle the more accurately the nearer their mutual distance approaches a right angle; to which the further reason is added, that for the equatorial stars not only the observed transit S, but also the assumed known right ascension α, are liable to the smallest absolute errors.

On closer consideration, however, one perceives that although the latter reason here may certainly have some weight, on the other hand the geometrical consideration is not here in place; that is to say, the two points that determine the meridian, *i. e.*, the pole and zenith, have, at every place, a given distance for each other which we cannot alter, and upon which, of course, directly depends the accuracy with which the plane of the meridian can in any way be determined. But the determination of absolute time consists, principally, in the perception of that part of the heavens with which the zenith of the place of observation coincides at a given instant, and will, therefore, be most directly attained by the observation of the transits of zenith stars. This is already shown in that for zenith stars the influence of the azimuth of the instrument disappears. Our formula makes the relations thus brought into consideration still plainer, and, if the necessary data were known with sufficient accuracy, would even allow a numerical estimation of the probable errors corresponding to the different cases. But also, independent of this, we see that on account of the factor $\operatorname{tg}\varphi - \operatorname{tg}\delta$, the influence of an error in the angle τ and in the adopted value of the quantity f disappears in the zenith, whereby it deserves to be particularly mentioned that the latter will be of special importance in the application of a movable thread. As to error in the inclination and in the error of collimation, the former influence is constant for all stars; the latter is greater the further the star is removed from the pole. It is, therefore, as above already remarked, only the reliability in the determination of D which diminishes with increasing inclination. But the principal law of this diminution is known to be very dependent upon the peculiarity of the observer, and the diminution, in general, first becomes noticeable at great declinations, so that finally results a very decided advantage for the zenith stars. This advantage will now be increased by the circumstance that for the observation in the zenith, in both positions of the instrument, the same portions of the pivots are used; their irregularities, therefore, are completely eliminated. On the other hand, it is to be allowed that the greater perfectness of the instrument in and of itself, and more especially the shortening of the duration of the observation, essentially diminishes all these different sources of error, and therefore gives a relatively greater weight to the error of the quantity D, which is not affected by them.

On taking a general view, therefore, the opinion appears authorized that zenith stars certainly have an advantage, but that, in our method, this is less important than in the establishment in the meridian; for in this latter the observation of zenith stars offers the only means of freeing ourselves from the unwarranted assumption of the invariability in azimuth, while in respect to the inclination, the possibly-existing changes at any moment can be recognized by means of the level. Such a greater freedom in the choice of the stars to be observed constitutes, if I am not mistaken, a further not unimportant advantage of our method.

13. It now, finally, remains to mention, with a few words, the reduction that is necessary for the Time Star, if this is observed on more than one thread. Because of the exhaustive treatment that this subject has received in different places, especially by Hansen, it suffices here to merely give the formulæ for computation. The time, t, that a star whose declination is δ needs in order to pass from any great circle of the sphere to the parallel circle, distant f therefrom, will be expressed with an accuracy entirely sufficient for our purposes by the formula:

$$t = f\sqrt{\sec(\delta + n) \,.\, \sec(\delta - n)},$$

where n denotes, as previously, the distance at which the great circle passes by the pole. A knowledge of n, sufficient for this reduction, is always at hand; if not otherwise, then it is offered by our computation itself, since for the reduction of the angle τ, the simple passage of the Time Star through the middle thread, or if this, perhaps, is not observed, a preliminary reduction of the side thread, with the factor $\sec\delta$, will suffice, without the least prejudice to the accuracy of the computation; and, in general, the exact reduction has only an importance if the observed threads do not lie symmetrical with respect to the middle thread.

14. It seems proper, in conclusion, to demonstrate, by some numerical

example, the accuracy and convenience of the above-given formula for computation. As a first example, I choose that given by Hansen at the conclusion of the memoir in No. 199 of the Astronomische Nachrichten. The data, according to the notation adopted by us, are as follows:

	h.	m.	s.		°	′	″
$\alpha =$	9	59	18.86	$\delta =$	12	47	33.6
$\alpha' =$	18	27	22.5	$\delta' =$	86	35	19.9

The Time Star was observed on three threads, the Pole Star only on the third. I give here the individual transits, together with the corresponding thread intervals:

Time Star. h.	m.	s.	Thread interval. s.	Pole Star. h.	m.	s.
10	51	47.7	+ 39.50			
	52	28.2				
	53	7.5	— 38.30	11	5	51 = S.

Finally,

$$\varphi = 50° \, 56' \, 0''; \quad b = -3''.4; \quad c = -4^s.70 = -70''.6.$$

We content ourselves at first in reference to S with the observation of the middle thread, without considering the two others, since we will not yet reduce them exactly; thus we have

	h.	m.	s.
$S - \alpha = D =$	0	53	9.34
$S' - \alpha' = D' =$	16	38	28.5
$D' - D =$	15	45	19.16
$\tau =$	236°	19′	47″.

We will now at first compute according to the exact formulæ of article 9, in order to test thereby the succeeding computation according to the approximate formulæ. I use six-figure logarithms in this in order that nowhere, by reason of the length of the computation, can the hundredths of a second of time be rendered doubtful through accumulation of errors in the last place; but I give the computation, not completely, but only its principal steps.

The formulæ I give:

		°	′	″		°	′	″
	$\xi =$	—2	53	25.6	$d =$	+10	53	10.9
thence from II follows:	$\eta =$	—	10	43.3				
therefore,	$\xi + \eta =$	—3	4	8.9				

And with this, according to III:

	$x =$	—0	39	39.87	$n =$	— 2	59	50.0
finally, from IV:	$m =$	—3	41	59.69				
therefore,	$x - m =$	+3	2	19.82				

With the n, thus found, follows

$$\log \sqrt{\sec(\delta + n) \,.\, \sec(\delta - n)} = 0.01154,$$

and with this the reduction for the time stars to the middle thread becomes $+ 40^s.56$ and $- 39^s.33$. The three transits through the middle thread are, therefore,

		h.	m.	s.
		10	52	28.56
				28.2
				28.17
therefore, more accurately,	$S=$	10	52	28.21
and	$D=$	0	53	9.35
But we have	$\frac{x-m}{15}=$		12	9.32
therefore, finally,	$u=$		—41	0.03

For the computation, according to the formulæ of article 11, now to be given, I borrow from the foregoing the value of

		°	′
	$d=$	10	53.2
Then is	$90^\circ-d=z'+z=$	79	6.8
further,	$\varphi-\delta=z=$	38	8.4
whence,	$z'=$	40	58.4

which I assume to have been read from the finding circle or taken from an ephemeris of the Pole Star. As above remarked, these quantities are only necessary in the computation of the coefficients C and F. In respect to the former I remark, that, in general, not its logarithm but the number itself will be first used, at least in case the error of collimation is not assumed as known, but is first to be deduced from the observations themselves. In such cases the above-given expression for C is quite convenient, especially if we observe that one portion of it recurs again in F. In other cases, as therefore in the present, it is more convenient to change it into

$$C=\sec\varphi\cos\left(\frac{z'-z}{2}\right)\sec\left(\frac{z'+z}{2}\right).$$

As regards the term $B\,b$, we should not hesitate to connect its computation with the equally necessary conversion of the level divisions, as directly given, into the corresponding value in arc; that is to say, we seek not

$$b=-3''.4,\ \text{but at once}\ B\,b=\frac{b}{15}\cdot\sec\varphi=-0^{s}.360.$$

From the given values of z' and z follows

$$\frac{z'+z}{2}=\sigma=39^\circ\ 33''.4$$

$$\frac{z'-z}{2}=\varDelta=\ 1\ \ 25.0$$

and, with these, the complete computation of the terms $C\,c$ and $F\,f$ become the following:

$\log\cos\varDelta=9.99987$	$\log C_0=0.3133$	$\log F_0=9.99909$
$\log\sec\sigma=0.11295$	$\log c=0.6721n$	$\log f=1.58320n$
$\log\sec\varphi=0.20050$	$\log C_0 c=0.9854n$	$\log F_0 f=1.58229n$
$\log\sin z=9.79070$	$q\,.\log\sec m=0.0015$	$k\log\sec m=0.00227$
$\log\sec d=0.00789$	$\log C\,c=0.9869n$	$\log F\,f=1.58456n$

The corrections dependent upon sec m have been later introduced, after obtaining the values of μ and m. Their omission would have introduced an error of only $0^s.033$ in Cc but of $0^s.200$ in Ff.

I have thus previously executed the computation of the terms Cc and Ff, because, as above remarked, in the actual application of our method, I suppose the factors C and F to be, once for all, computed and brought into a table. I will hereafter give a portion of such a table. The following is the computation that is now to be independently executed for each separate observation:

		°	′	″
$\log \operatorname{tg} \varphi = 0.090598$	$x_0 = -$		38	25.90
$\log \operatorname{tg} \delta = 9.356140$	$m_0 = -$	3	28	38.63
$\log \operatorname{cotg} \delta' = 8.775291$	$x_0 - m_0 = +$	2	50	12.7
$\log$ denominator $= -3247$				
			m.	s.
$\log \sin \tau = 9.920250n$	$D = +$		53	9.35
$\log \lambda = 8.131431$	$Bb = -$			0.360
$\log \cos \tau = 9.74383n$	$Cc = -$			9.703
Argum $= 2.12474\sigma$	$Ff = -$			38.420
$\log \mu = 0.734458$			+ 52	20.83
$\log \operatorname{tg} x_0 = 8.048434n$	$\frac{x_0 - m_0}{15} = +$		11	20.85
$\log \cos x_0 = -27$	$u = -$		40	59.98
$\log \sin m_0 = 8.782865n$				

that is to say, differing by $0^s.05$ from the result of the exact computation. Now, it is not difficult to see where the cause of this discordance is to be sought. By reason of the magnitude of the term Ff, it is certainly not surprising that the terms of the higher order neglected therein have not been entirely evanescent. But instead of developing these, it is certainly much more convenient, in such a case, to execute the computation rigorously as regards f, and the thereby ~~measured labor~~ proves itself to be so slight, especially in consideration that the Ff now entirely falls out, that it seems advisable always to choose this solution if the factor F is not already elsewhere given. The system of equations will be as follows:

$$\operatorname{tg} \xi = \frac{\sec \delta \operatorname{cotg} \delta' \sin \tau}{1 - \operatorname{tg} \delta \operatorname{cotg} \delta' \cos \tau}$$

$$\sin \eta = \frac{\sin f}{\sin (z' + z)}$$

$$\operatorname{tg} x_1 = \sin \delta \operatorname{tg} (\xi + \eta)$$

$$\sin m_1 = \cos \delta \operatorname{tg} (\xi + \eta) \operatorname{tg} \varphi \cos x_1$$

$$u = \left(\frac{x_1 - m_1}{15}\right) - (D + Bb + Cc)$$

These being applied to our example, the following computation results, which I here give without any omission:

			°	′	″
$\log \operatorname{tg} \delta = 9.356140$	$\log f = 2.75929n$	$\xi = -$	2	53	25.57
$\log \operatorname{cotg} \delta' = 8.775291$	$\log \sin (z' + z) = 9.99211$	$\eta = -$		9	45.03
$\log \sec \delta = 0.010916$	$\log \eta = 2.76718n$	$\xi + \eta = -$	3	3	10.6
$\log \operatorname{tg} \delta \operatorname{cotg} \delta' = 8.13143$	$\log \sin \delta = 9.345224$				

			°	′	″
$\log\cos\tau = 9.74383n$	$\log \mathrm{tg}\,(\xi+\eta) = 8.727007n$	$x_1 = -$		40	35.77
Argum. $= 2.12474\sigma$	$\log\cos\delta = 9.989084$	$m_1 = -$	3	40	24.58
$\log\sec\delta\,\mathrm{cotg}\,\delta' = 8.786207$	$\log \mathrm{tg}\,\phi = 0.090598$	$x_1 - m_1 = +$	2	59	48.81
				m.	s.
$\log\sin\tau = 9.920250n$	$\log\cos x_1 = \quad -30$	$\frac{x_1 - m_1}{15} = +$		11	59.254
log denominator $= \quad -3247$	$\log \mathrm{tg}\,x_1 = 8.072231n$	$D + Bb + Cc = +$		52	59.287
$\log \mathrm{tg}\,\xi = 8.703210n$	$\log\sin m_1 = 8.806659n$	$u = -$		41	0.03

15. The previous example is treated with abundant fullness, in order to leave no doubts as to the way to be pursued in all cases that can occur. It is, however, in fact more unfavorably conditioned than is necessary to assume in actual practice. First, α Ursæ Minoris should be always observed; and, secondly, a collimation error of nearly five seconds of time is an aggravation that can be avoided with s ome attention. The example is, also, evidently only a fictitious case. I will now, by the computation of an observation actually made, and noways specially favorable, show the process that we are accustomed to follow in the treatment of this problem. To this end we have computed, once for all, the coefficients C and F, for our latitude 59° 46′ 20″, for a number of the more frequently observed stars, under the assumption that the Time Star is observed some six minutes after the Pole Star, which corresponds very nearly to the interval that occurs in the actual observations. In the example here to be considered the following come into use:

	C	log C	log F
β Draconis - - - - -	2.060	0.3139	9.6151
γ Draconis - - - - -	2.069	0.3158	9.6576
α Lyræ - - - - - -	2.201	0.3427	9.9613
ζ Aquilæ - - - - - -	2.507	0.3997	0.1688

In reference to the computation itself, I make now, further, the following remarks:

For $\delta = 0$ the coefficient μ attains an infinite value; since, however, at the same time λ becomes $= 0$, our equations in this case are

$$\mathrm{tg}\, x_0 = 0 \qquad \sin m_0 = \mathrm{tg}\,\varphi\;\mathrm{cotg}\,\delta'\,\sin\tau.$$

But this indicates, that in order to be able, in all cases, to conduct the computation according to quite similar ways, whereby it is known that in the actual application no slight relief is attained, we must subject our formulæ still to a small change. We put

$$\lambda \,.\, \mu = \mathrm{tg}\,\varphi\;\mathrm{cotg}\,\delta' = \nu$$

and

$$\frac{\sin\tau}{1 - \lambda\cos\tau} = \rho$$

then will

$$\mathrm{tg}\, x_0 = \lambda \,.\, \rho$$

and

$$\sin m_0 = \nu \,.\, \rho \,.\, \cos x_0\,;$$

and the further advantage is here connected that the factor ν can be considered as constant for all observations of the same evening.

Since we always observe the Pole Star, the entire angle to be computed does not reach 3° even for a latitude of 60°. For such angles as is well known, the excellent five-figure logarithmic tables of Westphal, by the use of the small auxiliary tables headed "Corr.," in connection with the logarithms of the numbers, make the use of the trigonometric tables quite unnecessary; and it may be here remarked that we have, with great regret, missed this addition in the otherwise perfect tables of Bremiker. On this account, principally, we confine ourselves to five decimals in this computation, although thereby, under some circumstances, the hundredths of a second of time certainly become unsafe; but this can be considered as no noticeable diminution of accuracy in the observation of the transit of a single star.

As to the following observations, they have been made by a talented and industrious young officer, Mr. Koverski, who at present, as a pupil of the military academy, takes part in the two-year practical course at Poulkova; and, indeed, at that time he had had an experience of only the first few weeks in such observations. The instrument is an Ertel portable transit instrument of larger dimensions, but with the above-mentioned erroneous position of the clamp screw. The reticule consists of nine threads, and we are accustomed to indicate these with the numbers I to IX in fixed order, namely, according to their increasing distances from the ocular end of the axis. The otherwise customary notation, according to the order in which the observed stars pass through them, becomes ambiguous when the Pole Star stands very near its elongation. Small marks upon the threads themselves make a confusion impossible in our method of numbering; and this is important, since in the observation of the Pole Star upon only one thread an error, in this respect, would be very dangerous. In the use of a moveable thread, the numbers upon the micrometer are also quite as safe a means of indicating the fixed threads. The distances from the middle thread, that have come into use, are the following:

$$\text{VI} = 5^s.731, \qquad \text{VII} = 17^s.701.$$

These refer to the first of our examples, in which the observations in both positions are made in the same azimuth; while in the second, the Pole Star was too near the elongation, and therefore by changing the azimuth of the instrument, it was each time observed upon the middle thread. The following table contains the entire computation, exactly in the form in which we always conduct it. It needs certainly scarcely any further explanation, except that the factor sin 1″ is omitted on the lines distinguished by the angular brackets. The star positions are those of the British Nautical Almanac. The constants for the evening are:

$$\alpha' = 1^h\ 9^m\ 43^s$$
$$\log \cot \delta' = 8.39474$$
$$\log \nu = 8.62933$$

Date.	1863, July 29.		1863, July 29.	
Position.	East.	West.	West.	East.
Time star.	β Draconis.	γ Draconis.	α Lyræ.	ζ Aquilæ.
	h. m. s.	h. m. s.	h. m. s.	h. m. s.
$S' + \gamma$	17 25 4 VI	17 55 2 VII	18 34 44 M	19 4 20 M
S	17 32 44.55	17 59 9.69	18 40 51.39	19 10 45.44
a	17 27 23.05	17 53 28.54	18 32 21.35	18 59 10.47
D′	16 15 21	16 45 19	17 25 1	17 54 37
D	0 5 21.50	0 5 41.15	0 8 30.04	0 11 34.97
B b	— 0.50	— 0.18	— 0.15	— 0.27
F f	— 2.36	+ 8.05	0.00	0.00
	h. m. s.	h. m. s.	h. m. s.	h. m. s.
D′ — D	16 9 59.5	16 39 37.8	17 16 31.0	17 43 2.0
	° ′	° ′	° ′	° ′
τ	242 29.9	249 54.5	259 7.8	260 45.5
δ	52 24.5	51.30.6	38 39.8	13 40.0
log tg δ	0.11358	0.09955	9.90314	9.38589
log sin τ	9.94792n	9.97274n	9.99214n	9.99881n
log denominator	— 642	— 463	— 162	— 19
Argument	1.82725σ	1.96975σ	2.42662σ	3.3504σ
log cos τ	9.66443n	9.53596n	9.27550n	8.8690n
log λ	8.50832	8.49429	8.29788	7.78063
[log ρ]	5.25593n	5.28254n	5.30495n	5.31305n
log cos x_0	— 17	— 18	— 8	— 1
[log tg x_0]	3.76425n	3.77683n	3.60283n	3.09368n
[log sin m_0]	3.88509n	3.91169n	3.93420n	3.94237n
	° ′ ″	° ′ ″	° ′ ″	° ′ ″
x_0	— 1 36 49.5	— 1 39 40.1	— 1 6 46.6	— 0 20 40.7
m_0	— 2 7 57.0	— 2 16 2.0	— 2 23 16.6	— 2 26 0.0
$x_0 - m_0$	+ 0 31 7.5	+ 0 36 21.9	+ 1 16 30.0	+ 2 5 19.3
	m. s.	m. s.	m. s.	m. s.
$\frac{1}{15}(x_0 - m_0)$	+ 2 4.50	+ 2 25.46	+ 5 6.00	+ 8 21.29
D + B b + F f	+ 5 18.64	+ 5 49.02	+ 8 29.89	+ 11 34.70
u + C c	— 3 14.14	— 3 23.56	— 3 23.89	— 3 13.41
Clock rate	— 0.04	+ 0.04	— 0.04	+ 0.04

The chronometer gained 4ˢ.0 in twenty-four hours' sidereal time, and corresponding to this rate are the reductions to the mean moment of each pair of determinations, as given in the last line of the individual clock corrections. Such a reduction is necessary to the exact deduction of the collimation error. If we indicate this by $+c$ for position west, and therefore $-c$ for the position east, we have now, by means of the coefficients C, the following determinations:

	m. s.	m. s.
	$u - 2.060 \,.\, c = -3$ 14.18	$u + 2.201 \,.\, c = -3$ 23.93
	$u + 2.069 \,.\, c = -3$ 23.52	$u - 2.507 \,.\, c = -3$ 13.37
therefore,	$+ 4.129 \,.\, c = \quad -9.34$	$+ 4.708 \,.\, c = - \quad 10.56$
and thence,	$c = \quad -2.262$	$c = - \quad 2.243$
	$u = -3$ 18.84	$u = -3$ 18.99
	for 17ʰ 46ᵐ	for 18ʰ 56ᵐ

The agreement of the two values of c is thoroughly satisfactory, and the change in value of u corresponds, within 0ˢ.04, to the above-given rate of the chronometer.

TABLE I.

FINDING EPHEMERIS OF POLARIS.—$\delta = 88° 37'$.

Hour angle.	$\phi=+20°$. Z.	$\phi=+20°$. A.	$\phi=+30°$. Z.	$\phi=+30°$. A.	$\phi=+40°$. Z.	$\phi=+40°$. A.	$\phi=+50°$. Z.	$\phi=+50°$. A.	$\phi=+60°$. Z.	$\phi=+60°$. A.	$\phi=+70°$. Z.	$\phi=+70°$. A.	Hour angle.
h.	° ′	° ′	° ′	° ′	° ′	° ′	° ′	° ′	° ′	° ′	° ′	° ′	h.
0	68 37	0 0	58 36	0 0	48 37	0 0	38 36	0 0	28 37	0 0	18 37	0 0	24
1	40	23	39	24	40	28	39	34	40	0 45	39	1 7	23
2	48	44	48	48	48	55	47	1 6	49	1 26	48	2 9	22
3	69 1	1 3	59 2	1 8	49 1	1 18	39 1	33	29 3	2 1	19 2	3 0	21
4	19	17	18	24	18	35	18	53	20	27	20	3 37	20
5	39	26	39	33	39	45	39	2 6	40	42	40	3 58	19
6	70 0	1 28	60 0	36	50 0	48	40 0	9	30 2	46	20 1	4 3	18
7	21	25	21	32	21	44	21	2 4	20	39	24	3 50	17
8	41	16	42	22	42	34	42	1 50	39	21	42	3 23	16
9	59	1 2	59	1 7	59	1 16	59	30	30 56	1 54	20 59	2 44	15
10	71 12	0 44	61 12	0 47	51 12	0 53	41 19	1 3	31 10	1 20	21 11	1 55	14
11	20	23	21	24	21	0 26	21	0 32	20	0 41	20	0 59	13
12	23	0 0	24	0 0	23	0 0	24	0 0	23	0 0	22	0 0	12

A is to be added to 180° to obtain the azimuth when the hour angle is $>12^h$.
A is to be subtracted from 180° to obtain the azimuth when the hour angle is $<12^h$.

TABLE II.

COEFFICIENT OF AZIMUTHAL CHANGE FOR A UNIT OF HOUR ANGLE.

δ	$\phi=+20$.	30.	40.	50.	60.	70.	δ
°							°
− 50	+ 0.68	. .	. .	. .	. .	. .	− 50
40	0.81	+ 0.82	. .	. .	. .	. .	40
30	1.13	1.00	+ 0.92	. .	. .	. .	30
20	1.46	1.23	1.08	+ 1.00	. .	. .	20
− 10	1.97	1.53	1.28	1.14	+ 1.05	. .	10
0	2.92	2.00	1.56	1.30	1.15	+ 1.06	− 0
+ 10	+ 5.67	2.68	1.97	1.53	1.28	1.14	+ 10
20	∞	+ 5.41	2.75	1.88	1.46	1.23	20
30	− 4.99	∞	+ 4.99	2.53	1.73	1.35	30
40	2.14	− 4.41	∞	+ 4.41	2.24	1.53	40
50	1.28	1.88	− 3.70	∞	+ 3.70	1.88	50
60	0.78	1.00	1.46	− 2.88	∞	+ 2.88	60
70	− 0.45	− 0.53	− 0.68	− 1.00	− 1.97	∞	70
110	. .	+ 0.35	+ 0.36	+ 0.40	+ 0.45	+ 0.53	110
120	. .	. .	+ 0.51	0.53	0.58	0.65	120
130	. .	. .	. .	+ 0.65	0.68	0.74	130
140	. .	. .	. .	. .	+ 0.78	0.82	140
+ 150	. .	. .	. .	. .	. .	+ 0.88	+ 150

TABLE III.

Angle.		S.	T.	Angle.		S.	T.
″	′			″	° ′		
0	0	0	0	3600	1 0	22	44
60	1	0	0	3660	1 1	23	45
120	2	0	0	3720	1 2	24	47
180	3	0	0	3780	1 3	24	48
240	4	0	0	3840	1 4	25	50
300	5	0	0	3900	1 5	26	52
360	6	0	0	3960	1 6	27	53
420	7	0	0	4020	1 7	28	55
480	8	1	1	4080	1 8	28	57
540	9	1	1	4140	1 9	29	58
600	10	1	1	4200	1 10	30	60
660	11	1	1	4260	1 11	31	62
720	12	1	2	4320	1 12	32	63
780	13	1	2	4380	1 13	33	65
840	14	1	2	4440	1 14	34	67
900	15	2	3	4500	1 15	35	69
960	16	2	3	4560	1 16	36	71
1020	17	2	3	4620	1 17	36	73
1080	18	2	4	4680	1 18	37	74
1140	19	2	4	4740	1 19	38	76
1200	20	3	5	4800	1 20	39	78
1260	21	3	5	4860	1 21	40	80
1320	22	3	6	4920	1 22	41	82
1380	23	3	6	4980	1 23	42	84
1440	24	4	7	5040	1 24	43	86
1500	25	4	8	5100	1 25	44	88
1560	26	4	8	5160	1 26	45	90
1620	27	5	9	5220	1 27	46	93
1680	28	5	9	5280	1 28	48	95
1740	29	5	10	5340	1 29	49	97
1800	30	6	11	5400	1 30	50	99
1860	31	6	12	5460	1 31	51	101
1920	32	6	12	5520	1 32	52	104
1980	33	7	13	5580	1 33	53	106
2040	34	7	14	5640	1 34	54	108
2100	35	8	15	5700	1 35	55	110
2160	36	8	16	5760	1 36	57	113
2220	37	9	17	5820	1 37	58	115
2280	38	9	18	5880	1 38	59	118
2340	39	9	18	5940	1 39	60	120
2400	40	10	19	6000	1 40	61	122
2460	41	10	20	6060	1 41	63	125
2520	42	11	21	6120	1 42	64	127
2580	43	11	23	6180	1 43	65	130
2640	44	12	24	6240	1 44	66	132
2700	45	13	25	6300	1 45	68	135
2760	46	13	26	6360	1 46	69	138
2820	47	14	27	6420	1 47	70	140
2880	48	14	28	6480	1 48	72	143
2940	49	15	29	6540	1 49	73	145
3000	50	15	30	6600	1 50	74	148
3060	51	16	32	6660	1 51	76	151
3120	52	17	33	6720	1 52	77	154
3180	53	17	34	6780	1 53	78	156
3240	54	18	36	6840	1 54	80	159
3300	55	19	37	6900	1 55	81	162
3360	56	19	38	6960	1 56	83	165
3420	57	20	40	7020	1 57	84	168
3480	58	21	41	7080	1 58	85	170
3540	59	21	43	7140	1 59	87	173
3600	60	22	44	7200	1 60	88	176

TABLE III.—Continued.

Angle.		S.	T.	Angle.		S.	T.
″	° ′			″	° ′		
7200	2 0	88	176	10800	3 0	199	397
7260	2 1	90	179	10860	3 1	201	401
7320	2 2	91	182	10920	3 2	203	406
7380	2 3	93	185	10980	3 3	205	411
7440	2 4	94	188	11040	3 4	207	415
7500	2 5	96	191	11100	3 5	21[illegible]	419
7560	2 6	97	194	11160	3 6	212	424
7620	2 7	99	198	11220	3 7	214	428
7680	2 8	100	201	11280	3 8	217	433
7740	2 9	102	204	11340	3 9	219	438
7800	2 10	104	207	11400	3 10	221	442
7860	2 11	105	210	11460	3 11	223	447
7920	2 12	107	213	11520	3 12	227	452
7980	2 13	108	217	11580	3 13	229	456
8040	2 14	110	220	11640	3 14	231	461
8100	2 15	112	223	11700	3 15	233	466
8160	2 16	113	227	11760	3 16	235	471
8220	2 17	115	230	11820	3 17	237	476
8280	2 18	117	233	11880	3 18	240	481
8340	2 19	118	237	11940	3 19	242	486
8400	2 20	120	240	12000	3 20	245	490
8460	2 21	122	243	12060	3 21	247	495
8520	2 22	124	247	12120	3 22	250	500
8580	2 23	125	250	12180	3 23	252	505
8640	2 24	127	254	12240	3 24	255	510
8700	2 25	129	258	12300	3 25	257	515
8760	2 26	131	261	12360	3 26	260	521
8820	2 27	132	265	12420	3 27	263	525
8880	2 28	134	268	12480	3 28	266	530
8940	2 29	136	272	12540	3 29	268	535
9000	2 30	138	276	12600	3 30	271	540
9060	2 31	140	279	12660	3 31	273	545
9120	2 32	142	283	12720	3 32	275	551
9180	2 33	144	287	12780	3 33	278	556
9240	2 34	145	291	12840	2 34	281	561
9300	2 35	147	294	12900	3 35	284	566
9360	2 36	149	298	12960	3 36	286	572
9420	2 37	151	302	13020	3 37	289	577
9480	2 38	153	306	13080	3 38	291	582
9540	2 39	155	310	13140	3 39	294	588
9600	2 40	157	314	13200	3 40	297	593
9660	2 41	159	318	13260	3 41	300	598
9720	2 42	161	322	13320	3 42	302	604
9780	2 43	163	325	13380	3 43	305	610
9840	2 44	165	330	13440	3 44	307	615
9900	2 45	167	334	13500	3 45	310	621
9960	2 46	169	338	13560	3 46	313	626
10020	2 47	171	342	13620	3 47	316	632
10080	2 48	173	346	13680	3 48	318	637
10140	2 49	175	350	13740	3 49	321	643
10200	2 50	177	354	13800	3 50	324	649
10260	2 51	179	358	13860	3 51	327	654
10320	2 52	181	362	13920	3 52	330	660
10380	2 53	183	367	13980	3 53	333	666
10440	2 54	186	371	14040	3 54	335	671
10500	2 55	188	375	14100	3 55	338	677
10560	2 56	190	380	14160	3 56	341	683
10620	2 57	192	384	14220	3 57	345	688
10680	2 58	194	388	14280	3 58	347	695
10740	2 59	196	393	14340	3 59	350	701
10800	2 60	199	397	14400	3 60	352	707

APPENDIX AND TABLES,

BY THE TRANSLATOR.

APPENDIX.

The memoir here presented to the reader's notice was published in 1863 by the Imperial Central Observatory at Poulkova, (near St. Petersburg,) after the methods herein given had been carefully elaborated by their author and often tested in the course of his experience as senior astronomer at that institution and as professor of geodesy to the Imperial Military Academy.

Until 1864 the practical application of Döllen's method was seriously embarrassed, inasmuch as no instrument specially proper to this class of observations had been constructed, and the use of "universal instruments" and Hansen's formulæ could only be said to have given promise of what would be achieved with a suitable instrument. In the spring of this year, however, there were finished by Brauer, the then mechanician of the observatory, three portable transits, two of which were designed for the immediate use of the observers engaged in determining the longitudes of points on the Valentia-Orsk parallel; the third, soon after its completion, passed into the hands of Professor Schidloffski, director of the observatory at Kieff, who took it with him on returning home in August from the quarter-century inauguration anniversary at Poulkova.

This latter instrument it was that, having been during the summer at the disposal of Mr. Döllen, would, as he hoped, enable him to carry out such special studies as would decide definitively upon the advantages resulting from its use, the instrument having been made by Brauer, under his own supervision, and with direct reference to the convenient application of his method for the determination of time.

But the series of observations then begun, unfortunately, could not be completed, and they still remain unpublished, and probably not fully discussed—a delay necessitated by the pressure of imperative duties consequent upon Döllen's appointment *ad interim* to the directorship of the observatory, and an illness that for months threatened his life, and finally compelled him to a year's absence from Poulkova.

In the summer of 1864 Messrs. Thiele and Zylinski, engaged on the above-mentioned arc of longitude, being in England, their instruments were brought so favorably to the notice of the astronomer royal that Professor Airy ordered from Brauer a similar portable transit for the Greenwich Observatory. This was finished in the winter of 1865–'66, and the accuracy of its construction was sufficiently tested during a few clear days in the following spring, the writer having himself improved the opportunity of examining the cylindricity of its pivots by means of the microscope apparatus supplied with it.

Simultaneously with the construction of the latter instrument Brauer took in hand two others, (Nos. V and VI,) which he was not then able to complete, owing to his other engagements and the removal of his workshop to St. Petersburg, but which he has promised shall be even superior to their predecessors, and which, if not now finished, certainly can be within a few months.

THE BRAUER PORTABLE TRANSIT.

The Brauer portable extra-meridional transit instrument consists, as to its chief features, of an objective of 2.5 to 3.0 inches aperture and 30 to 36 inches focal length, bearing a power of 150. The ocular being at one end of the axis of revolution, the light for the field illumination enters through the opposite pivot; the total reflection prism within the central cube admits of proper adjustment. To the stationary reticule of five wires is added a pair of close double wires, moved by a fine micrometer screw. The hanging level need never be removed from its position on the pivots. An unusually expeditious reversing apparatus is always in place, and works with scarcely any possibility of detriment to the stability of azimuth or level. This apparatus is so contrived as not to interfere with nadir observations, made by means of a dish of mercury placed on the pier and below the base of the tripod. The three foot screws rest on corresponding blocks, one of which is itself moveable by a horizontal adjusting screw, allowing the instrument to be accurately placed at any azimuth within a range of six or seven degrees. Microscopes and levels for the examination of the pivots are also provided.

The excellence of these instruments depends so much upon the convenient arrangement and conscientious workmanship of all the parts, that actual use and the critical study of the results can alone (as it actually does) suffice to persuade one of their superior merits.

While the Brauer transit is unusually convenient for use in any vertical plane, it may, of course, be very advantageously mounted in the meridian; it is in this way that the International Commission, having the Valentia-Orsk arc in charge, have decided to employ it, preferring, it would seem, to sacrifice time rather than risk the introduction of unknown errors by the adoption of new methods in a work of such magnitude and importance. That they, in 1863, may have been justified in this discussion will be admitted, but their own, as well as Döllen's, subsequent experience has removed all serious objections to the methods proposed and advocated by the latter. On the other hand, the experience of Messrs. Thiele and Zylinski has brought out in stronger light a peculiar personal equation that had been long known to exist in the use of meridian transit instruments and of the vertical circle for time determinations. It is found, namely, that the personal equation varies (in eye and ear observations very decidedly, but far less so in eye and hand observations) with the direction of the motion of the star's image over the retina. This source of error is not eliminated by the reversion of the ordinary direct-vision transit in the Y's, but may be so if a reflecting eye-piece be properly employed.

On account of the great saving of time and labor afforded by the use of the Brauer transit and Döllen's formulæ, these especially commend themselves to travelers and to those who are placed in unfavorable atmospheric conditions; the extra-meridional method here developed must be considered indispensable when the stability of the transit in azimuth or the visibility of the appropriate stars cannot be relied upon for several hours, but may be for perhaps five or ten minutes.

It will often be found highly advantageous for the observer to apply any ordinary portable transit to observations of this class however inconvenient the instrumental arrangements may seem to be.

For perspicuity and the observer's convenience we take the liberty of appending the following review of Döllen's method, together with a few simple tables.

ORIENTATION.

Arrived at any new station whose latitude as well as our clock-chronometer correction are approximately known, we determine, first, the zenith point of the setting circle by observation of the nadir point or of a distant terrestrial object. Then a finding ephemeris (such as is given in the appended Table I) enables us to set at the proper zenith distance for Polaris, and to sweep in azimuth until that star is found. This, if it be right, or if the north point be accurately known, will require but a few minutes.

In the day-time it will be more convenient to raise the telescope by means of its swerving apparatus and sight upon the sun. The observed time of transit (T) and the zenith distance (z) of the sun's center give us his azimuth (A) and the clock correction (Δ T) by the following well-known formulæ:

Put

$$90^\circ - \delta = a \qquad s - a = \alpha$$
$$90^\circ - \varphi = b \qquad s - b = \beta$$
$$z = c \qquad s - c = \gamma$$
$$2s = a + b + c \qquad 2s = \alpha + \beta + \gamma$$

where the double computation of $2s$ serves as a check:

$$n = \sqrt{\frac{\sin\alpha \sin\beta \sin\gamma}{\sin s}}$$

$$\operatorname{tg} \tfrac{1}{2} A = \frac{n}{\sin\alpha}, \qquad \operatorname{tg} \tfrac{1}{2} p = \frac{n}{\sin\beta}, \qquad \operatorname{tg} \tfrac{1}{2} t = \frac{n}{\sin\gamma},$$

$$\operatorname{tg} \tfrac{1}{2} A \operatorname{tg} \tfrac{1}{2} p \operatorname{tg} \tfrac{1}{2} t = \frac{n}{\sin S}$$

where the last equation again serves as a check.

Knowing thus the sun's azimuth, A, and hour angle, t, we have

North point = observed azimuth $\pm$ A

Clock correction = T — (right ascension of sun $+ t$).

The ordinary portable transit offers no very convenient means for measuring azimuths or for turning the instrument around through any given azimuthal angle. In the Brauer instrument this may be easily effected by means of a simple graduation on the vertical face of the exterior rim of the circular base of the transit, or more elegantly by means of a graduation in the circular base of the reversing apparatus or the adjacent bed-plate.

If the instrument does not contain too much iron in the construction of its parts an attached compass needle will be found convenient.

Polaris being found in the field of view, the careful observer will, after leveling the axis, reverse upon that star, or upon a distant terrestrial object, and see that the collimation error is not too large.

The choice of a time star will, in the day time, depend upon its own brightness, and a general catalogue of all stars, to the third magnitude inclusive, will afford a number of convenient stars at any time or in any latitude. We have appended such a catalogue, compiled by ———, including all stars given, as of the third or brighter magnitudes, in the British Association Catalogue, or Argelander's Uranometria Nova, or in Taylor's General Catalogue; to these stars we have, however, preferred to affix Argelander's magnitudes, excepting for those south of —25° of declination; for the variable stars we have adopted the limiting magnitudes given by Schönfeld. These are all converted into the decimal system of notation, assuming Argelander's 1.2 and 2.1, &c., equal to 1.4 and 1.6, &c., of the decimal system. For night observations the brightness of the star is not always a matter of so much importance, though it will be conducive to greater uniformity of personal equation if stars of equal magnitude be habitually employed. On the other hand, as we may save some little labor by using the stars whose apparent places are given in the Annual Astronomical Ephemerides, we have therefore added to our list of 200 bright stars *all others* given in the American Astronomical Ephemeris, (A,) in the British Nautical Almanac, (B,) in the Berliner Jahrbuch, (J,) or in the Connaissance des Temps, (C,) and have indicated the occurrence of a star in either of these almanacs by an appropriate letter in the second column of the resulting catalogue of 368 stars.

It will be noticed that we thus obtain a sufficient number of points for use in observing moon culminations should the traveler have time or inclination to resort to that method of determining longitude.

METHODS OF OBSERVATION.

The course now to be pursued by the observer depends, to some extent, upon the manner in which he proposes to use his instrument for the accurate determination of his clock corrections, but chiefly upon the atmospheric conditions and the stars of his ephemeris. Either of the following methods may be adopted; for simplicity we shall, as usual, speak of the observer as being in the northern hemisphere:

I.

The instrument being moved in azimuth until Polaris is within the reticule, and by preference near the middle thread, we read the level in both positions of the axis, observe the transit of Polaris over one thread and that of a time star immediately before or afterwards, and again read the level in both positions. Then, supposing the inequality of the pivots and the collimation to be known, we have the material for determining the clock correction. The collimation had best be at once determined by observations of Polaris in both positions of the axis.

But this method ought never to be resorted to except in extreme necessity; it is always of importance to institute observations of the time stars as well as of Polaris in both positions of the axis, as in the following methods.

II.

Read the level; observe the time star, and then Polaris, over any thread, by preference the middle one, or, still better, make several bisections by means of the micrometer thread; read the level; reverse the axis; read the level; observe Polaris again, as also another time star.

The observations of the time stars may either precede or follow those of Polaris.

This method, in which the azimuth remains the same in the two positions of the axis, is specially applicable when the azimuthal motion of Polaris is rapid, and differs from the ordinary use of the transit only in that Polaris is observed at any hour angle whatever.

III.

Read the level; observe Polaris on the middle thread and a time star; read the level; reverse the axis; read the level; alter the azimuth slightly, so that Polaris will again cross the middle thread in a few seconds; read the level; observe Polaris on the middle thread and a time star; read the level; reverse the axis, and again read the level.

The last reversion and level reading may be omitted if we have no reason to doubt the stability of the instrument.

In this method, which is commonly the most convenient and expeditious both as regards the observations and the computations, we alter the azimuth only by a small amount, so as to overtake Polaris in its diurnal motion. This method is specially applicable when no micrometer is provided for the instrument.

IV.

By the azimuthal motion bring the extreme western thread of the reticule to Polaris; read the level; observe the time star and Polaris; read the level; reverse the axis; read the level; bring the same thread, now become the extreme eastern one, to Polaris: read the level; observe Polaris and *the same time star*; read the level; reverse the axis, and again read the level.

The last reversion and level-reading may be omitted if the mounting is sufficiently stable.

The fixed thread of the reticule may be advantageously replaced by the extreme micrometer thread, if such an one is provided in addition to the central thread. In this method we reverse upon Polaris in order not to alter the azimuth so much as to leave that star outside the reticule. It is, however, more expeditious and quite safe to observe Polaris before the time star; then reverse and set upon *the same time star*, alter the azimuth so as to again observe its transit, and, finally, observe Polaris, which will certainly be found in the field.

In this method the azimuth is altered by a large amount, not exceeding, however, the angular distance of the extreme wires of the reticule.

PREPARATORY COMPUTATIONS.

In preparing for ~~either~~ The first three of the preceding methods of observation we make use of a solution of the problem giving the positions of two stars and of the observer's zenith, required the moment at which these three points are in the same vertical plane. Instead of a rigorous solution, however, we may, by means of Table II, attain a sufficiently approximate determination of the required moment.

For a star in the meridian the movement in azimuth is, approximately,

$$dA = \frac{\cos\delta}{\sin(\varphi - \delta)} dt.$$

Table II gives the values of dA for $dt = 1$; its use will be apparent from the following example:

Example.—At latitude + 33°, and about 16^h of sidereal time, we desire to determine the clock correction; we select δ Ophiuchi and γ Herculis as appropriate time stars, whose places are

		h.	m.		°	′
No. 234, δ Ophiuchi; 3;	$\alpha =$	16	7.5	$\delta = -$	3	21
No. 237, γ Herculis; 3;	$\alpha =$	16	16.2	$\delta = +$	19	28

For these two moments the ephemeris of Polaris gives

	h. m.	h. m.	° ′	° ′ ° ′
At $\varphi = 33°$	at 16 7:	$t = 14$ 56;	$z = 58$ 0;	A = 1 5 = 181 5
	at 16 16:	$t = 15$ 5;	$z = 57$ 57;	A = 1 8 = 181 8

From Table II we find

$$\text{At } \varphi = 33° \begin{cases} \text{For } \delta \text{ Ophiuchi } \dfrac{d\,A}{d\,t} = +\,1.8 \\ \text{For } \gamma \text{ Herculis } \dfrac{d\,A}{d\,t} = +\,2.2 \end{cases}$$

Whence the hour-angle of δ Ophiuchi, when at the azimuth, $-1°\,5'$, is

$$t = -\frac{1°\,5'}{1.8} = -\,0°.60 = -\,2^m.4$$

and for γ Herculis, $t = -\dfrac{1°\,8'}{2.2} = -\,0°.51 = -\,2^m.1$

and the moments at which the time stars are to be observed will be, respectively, $16^h\,9^m.9$ and $16^h\,18^m.3$.

In setting for the time stars we may use meridian zenith distances, $\varphi - \delta$, but having brought the star between the horizontal wires, when once it has entered the field, we should, for accuracy and security, record the reading of the setting circle, and that, too, if possible, to the fraction of a minute.

In following method IV of observation we need to know the space of time within which the observation, reversal, &c., must be accomplished, or the change of azimuth that can be made without throwing Polaris too far from the middle wire.

We have, with sufficient accuracy,

$$\delta\,A = \frac{i}{\sin z}$$

where i = the equatorial distance of the extreme from the middle thread, and z is the zenith distance of the time star to be observed. Assuming $i = 1^m = 15'$, we obtain the $\delta\,A$ of the following tables:

z.	δ A.	z.	δ A.
°	° ′	°	° ′
10	1 26	30	0 30
15	0 58	40	0 23
20	0 44	50	0 20
25	0 36	60	0 18
30	0 30	70	0 16

The azimuthal change to be given to the transit is, evidently, double the A of the preceding table, since the same wire is to be observed first when on the extreme west, and again when on the extreme east.

Example.—At latitude 54°, at about $12^h\,35^m$ sidereal time, we propose to observe γ Virginis for clock correction. We have:

Polaris $z = -\,37°\,22'$ $A = +\,0°\,19'$ $t = 11^h\,24^m$ $\delta\,A = 0°\,25'$

γ Virginis $z = +\,54°\,44'$ $\dfrac{d\,A}{d\,t} = +\,1.24$

Whence, if 1^m is the equatorial distance of the extreme wire, the two settings must be between the azimuths

$$A_1 = +0^\circ 19' + 0^\circ 25' = +0^\circ 44', \text{ and } A_2 = +0^\circ 19' - 0^\circ 25' = -0^\circ 6';$$

and the time elapsed between the time star transits over the middle threads of the diaphragm will be

$$\frac{0^\circ 50'}{1.24} = \frac{3^m.33}{1.24} = 2^m.7,$$

leaving thus at least forty seconds for reading the level and setting circle, reversing, reading the level, and setting at the proper zenith distance and azimuth. This is, indeed, as unfavorable a case as need occur. By omitting to observe the time star on the extreme wires, we can materially increase the interval of forty seconds. In instruments properly constructed the level is always in place and its reading occupies but a few seconds.

METHODS OF COMPUTATION.—COLLIMATION.

In case the collimation is to be independently determined by observations of Polaris in reversed positions of the axis, we may use the following method for deducing the value of that quantity:

The several micrometric bisections being made in rapid succession in each position of the axis, it will be sufficiently exact to consider the mean reading of the micrometer as belonging to the mean of the observed times. We have, then, as also in case any two fixed wires are observed,

At time T_1, first position, Polaris distant from the middle wire - - i_1
At time T_2, reversed position, Polaris distant from the middle wire - i_2

the collimation c, and the distances i_1 and i_2, being always supposed to increase in the direction of the positive motion of the micrometer screw and to be expressed in seconds of time, being known by observations of equatorial and polar stars.

Now, the ordinary formulæ for azimuth instruments give,

$$\text{At time } T_1,\ A_1 = a_1 + b_1 \operatorname{cotg} Z_1 + (c + i_1) \operatorname{cosec} Z_1,$$
$$\text{At time } T_2,\ A_2 = a_2 + b_2 \operatorname{cotg} Z_2 + (c + i_2) \operatorname{cosec} Z_2.$$

Where A_1 and A_2 are the azimuths, Z_1, Z_2 the zenith distances of Polaris, $a_1 = a_o + \delta_a p$ and $a_2 = a_o + \delta'_a p$ the azimuths of the axis of revolution in the direct and reversed positions, and which will be identical if the pivots are of perfect form, (or $\delta_a p = 0$ and $\delta'_a p = 0$.) So, also, will the inclinations $b_1 = b_o + \Delta_1 p + \Delta_2 p$ and $b_2 = b_o + \Delta'_1 p + \Delta'_2 p$ be identical if the pivots are of equal diameters and perfect form, (or $\Delta_1 p = 0$; $\Delta'_1 p = 0$; $\Delta_2 p = 0$; $\Delta'_2 p = 0$.)

For Z_1 and Z_2 we may use the mean zenith distance, Z, taken from the finder, or the finding ephemeris for the mean moment, $\frac{1}{2}(T_1 + T_2) = T$.

As the observations are supposed to be made in rapid succession, we may assume

$$A_1 = A + \frac{dA}{dt}(T_1 - T),$$

$$A_2 = A + \frac{dA}{dt}(T_2 - T),$$

where

$$\frac{dA}{dt} = \cos\delta \frac{\cos Q}{\sin Z};$$

$2c = (i_2 - i_1) + (b_2 - b_1) \cos Z - (T_2 - T_1) \frac{dA}{dt} \cdot \sin Z.$

Q being the parallactic angle, which is given with sufficient accuracy by

$$\sin Q = \cos \varphi \frac{\sin (\alpha - T)}{\sin Z}.$$

There now results, for the collimation,

$$2c = (i_2 - i_1) \operatorname{cosec} Z + \frac{dA}{dt}(T_1 - T_2) + (b_2 - b_1) \cos Z;$$

c will be negative if the middle thread or micrometer zero is too far in the direction of positive motion.

TIME.

The deduction of the clock correction is to be effected by slightly different processes of computation, according as we have followed one or the other of the previous methods of observation. The formulæ given by Döllen may be systematically arranged as follows, using the following notation as adopted from the preceding memoir:

Pole Star.	Time Star.	
S'	S	the observed times of transit.
$u' = u + \gamma$	u	the clock corrections on sidereal time; γ the correction for clock rate during $S' - S$.
α'	α	the apparent right ascensions, corrected for diurnal aberration, if necessary.
δ'	δ	the apparent declinations.
$z' = \delta' - \varphi$	$z = \varphi - \delta$	the zenith distances, z' being always positive.
f		the interval for the thread on which Polaris is observed, (+ when west of the middle.)
	c	the error of collimation, (positive when the middle thread lies too far to the west.)
	b	the inclination, (positive for west end high.)
	a	the azimuth of the axis, (positive when its west end lies south of west point.)

COLLIMATION KNOWN.

One pole star and one time star, observed in either position of the axis; the inclination, (b,) the latitude, (φ,) and thread interval, (f,) known.

1.—*Rigorous solution.*

(See Döllen, § 14, first solution.)

(a.) Compute τ from

$$S' + \gamma - \alpha' = D'; \quad S - \alpha = D; \quad 15(D' - D) = \tau.$$

(b.) Compute $\sin d$, $\cos d$, and ξ from

$$\cos d \sin \xi = \cos \delta' \sin \tau$$
$$\cos d \cos \xi = \cos \delta \sin \delta' - \sin \delta \cos \delta' \cos \tau$$
$$\sin d = \sin \delta \sin \delta' + \cos \delta \cos \delta' \cos \tau.$$

(c.) Compute η from

$$\sin \eta = \frac{\sin (c + f) - \sin c \sin d}{\cos c \cos d}.$$

(*d.*) Compute n, x, and m from

$$\begin{aligned}\cos n \cos x &= +\cos c \cos(\xi+\eta)\\ \cos n \sin x &= -\cos\delta \sin c + \sin\delta\cos c \sin(\xi+\eta)\\ \sin n &= +\sin\delta\sin c + \cos\delta\cos c\sin(\xi+\eta)\\ \sin m &= \operatorname{tg} n \operatorname{tg}\varphi + \sin b \sec n \sec\varphi\end{aligned}$$

and deduce $\frac{1}{15}(x-m)$.

(*e.*) Reduce each observed thread to the middle one by the reductions

$$\tau = f\sqrt{\sec(\delta+n)\sec(\delta-n)}$$

and derive a new S from the mean of all, whence a new D is to be found.

This will rarely necessitate a correction to the $\frac{1}{15}(x-m)$ already computed.

(*f.*) With the corrected D compute

$$u=\frac{1}{15}(x-m)-\mathrm{D}$$

2.—*Approximate solution.*

When Ff is large, or the factor F is not given. (See Döllen, § 14, third solution.)

(*a.*) Compute τ as in (1 . *a*) from

$$\mathrm{S}'+\gamma-\alpha'=\mathrm{D}';\quad \mathrm{S}-\alpha=\mathrm{D};\quad 15(\mathrm{D}'-\mathrm{D})=\tau.$$

(*b.*) Compute ξ from

$$\operatorname{tg}\xi=\frac{\sec\delta\operatorname{cotg}\delta'\sin\tau}{1-\operatorname{tg}\delta\operatorname{cotg}\delta'\cos\tau}.$$

(*c.*) Compute η from

$$\sin\eta=\frac{\sin f}{\sin(z'+z)}.$$

(*d.*) Compute x_1 and m_1 from

$$\begin{aligned}\operatorname{tg} x_1 &= \sin\delta\operatorname{tg}(\xi+\eta)\\ \sin m_1 &= \cos\delta\operatorname{tg}(\xi+\eta)\cos x_1\operatorname{tg}\varphi.\end{aligned}$$

(*e.*) We now have

$$u=\frac{1}{15}(x_1-m_1)-\mathrm{D}-(\mathrm{B}b+\mathrm{C}c)$$

and there remains only to reduce all the transits of the time star to the middle thread, to correct D thereby, and to compute the terms $\mathrm{B}b+\mathrm{C}c$.

If the value of S, first used, be the mean of all threads systematically disposed about the mean thread, it will, generally, not require any correction for the quantity n.

(*f.*) The inclination, b, is given in level divisions, one of which has the value, p, expressed in seconds of time; whence we can easily compute

$$\mathrm{B}b=bp\sec\varphi.$$

(*g.*) The collimation, c, is expressed in seconds of time, and we have to compute

$$\begin{aligned}\mathrm{C}_0 &= \sec\varphi\cos\tfrac{1}{2}(z'-z)\sec\tfrac{1}{2}(z'+z)\\ q &= \operatorname{cosec}^2\varphi+(\operatorname{cotg}\varphi-\operatorname{tg}\delta)\operatorname{tg}\tfrac{1}{2}(z'-z)\\ \mathrm{C} &= \mathrm{C}_0(\sec m_1)^q;\end{aligned}$$

whence results $\mathrm{C}\,.\,c$.

It will, if c be moderately small, be always allowable to use C_0 instead of C.

3.—*Approximate solution.*

When Ff is not large or the factor F is given. (See Döllen, § 14, second solution.)

(*a.*) Compute τ, as before, from

$$S' + \gamma - a' = D'; \quad S - a = D; \quad 15\,(D' - D) = \tau;$$

where we shall not materially increase the probable error of our result by adding to the threads, observed symmetrically, any others reduced to the mean thread, by the ordinary formula,

$$f \sec \delta.$$

(*b.*) Compute λ, μ, ν, and ρ, from

$$\lambda = \text{tg}\,\delta \,\text{cotg}\,\delta'$$

$$\mu = \text{tg}\,\varphi \,\text{cotg}\,\delta \qquad\qquad \rho = \frac{\sin \tau}{1 - \lambda \cos \tau}.$$

$$\nu = \lambda\,\mu = \text{tg}\,\varphi \,\text{cotg}\,\delta'$$

(*c.*) Compute x_0 and m_0 from

$$\text{tg}\,x_0 = \lambda\,\rho, \quad \sin m_0 = \nu\,\rho \cos x_0.$$

(*d.*) We now have

$$u = \frac{1}{15}(x_0 - m_0) - D - (Bb + Cc + Ff)$$

and it remains to compute the three terms,

$$Bb + Cc + Ff.$$

(*e.*) As in (2.*f*) we have, for the inclination,

$$Bb = bp \sec \varphi.$$

(*f.*) For the collimation we have, as in (2.*g*),

$$C_0 = \sec \varphi \frac{\sin z' + \sin z}{\sin (z' + z)} = \sec \varphi \cos \tfrac{1}{2}(z' - z) \sec \tfrac{1}{2}(z' - z)$$

$$q = \text{cosec}^2 \varphi + \left(1 - \frac{1}{\mu}\right) \text{cotg}\,\varphi \,\text{tg}\,\tfrac{1}{2}(z' - z)$$

and $\qquad C = C_0 (\sec m_0)^q;$

where, as before, for all ordinary values of c, we may assume C equal to C_0.

(*g.*) The equatorial thread, or micrometer interval, f, being expressed in seconds of time, we have

$$F_0 = \sec \varphi \frac{\sin z}{\sin (z' + z)}$$

$$k = 1 + 2\,\text{cotg}^2 \varphi + \frac{1}{\mu}$$

and $\qquad F = F_0 (\sec m_0)^k.$

If F is not given by appropriate tables, then, for very small values of f, we may assume F equal to F_0.

COLLIMATION UNKNOWN.

Each of the three methods of computation previously given assumes the collimation (c) to be previously known, and is adapted to the independent deduction of a clock correction for each pair of stars observed in either position of the instrument. But it is evident that we have only to make a very slight change in the order of computation so as to deduce both collimation and clock correction from the observation of two pairs of stars, one in each position. It is this class of observations that is to be highly recommended and especially enjoined.

We may leave out of consideration the rigorous methods of computation and confine ourselves to the more generally useful approximate methods.

4.—*Approximate solution.*

When Ff is large, or the factor F not given. (See Döllen, § 15, first example.)

The azimuth may have remained the same and two different stars have been used for time stars, (as in the II method of observation,) or the azimuth may have been altered in order to observe the same time star in both positions, (as in the IV method of observation.) In either case we have S'_1 and S_1 for the observations of the Pole Star and Time Star in the first position, but S'_2 and S_2 for the observations in the reversed position, and we have to compute the observations of each position independently, according to the method 2 merely in article (2 . g), stopping at the computation of

$$C_0,\ q, \text{ and } C.$$

We thus obtain, by (2 . e), from the first time star,

$$u_1 \pm C_1 c = \frac{1}{15}(x_1 - m_1) - (D_1 + B b_1),$$

and from the second time star,

$$u_2 \mp C_2 c = \frac{1}{15}(x_2 - m_2) - (D_2 + B b_2).$$

Since each clock correction, u_1 and u_2, is supposed to hold for the moment of the observation of the time star, we reduce them (by adding the quantities γ_1 and γ_2, computed with an approximate clock rate) to a common moment, for which let u be the clock correction; we then have

$$u \pm C_1 c = \frac{1}{15}(x_1 - m_1) - (D_1 + B b_1) - \gamma_1 = O_1$$

$$u \mp C_2 c = \frac{1}{15}(x_2 - m_2) - (D_2 + B b_2) - \gamma_2 = O_2,$$

where the upper signs hold when the first position of the instrument is the normal one. The half sum and difference of O_1 and O_2 now give us the complete solution of our problem.

5.—*Approximate solution.*

When Ff is small, or F previously given. (See Döllen, § 15, second example.)

The azimuth may have remained the same and Polaris have been observed over side threads or the micrometer wire, (as in the II method of observation,) or the azimuth may have been changed and Polaris observed on the middle thread, (as in the III method of observation.)

We, in the first case, follow the method 3 with a modification precisely similar to that given in the previous method, 4, and we obtain, as there, from the values of $u_1 + \gamma_1 \pm C_1 c$ and of $u + \gamma_2 \pm C_2 c$ the correct u and c. (See Döllen, § 15, first example.)

In the second case, following method III, Ff is zero; the computation is made according to 2 and 4, but becomes simplified since $\eta = 0$. This is, probably, the best of all the methods of observation and computation. (See Döllen, § 15, second example.)

NOTE 1.—If we have pursued the IV method of observation, we may assume $C_1 = C_2$ without sensible error and obtain u without computing C or c. But it is always desirable not to omit the determination of the collimation.

NOTE 2.—We have now reviewed the five methods of computation, some one of which is specially appropriate to the use of the transit instrument, according to either of the four methods of observing previously given. The expert observer will not find it difficult so to arrange his observations that five-place logarithms may be used in the computations, and, as in the use of small angles the ordinary five-figure tables of logarithmic-trigonometric factors may be advantageously replaced by the tables of logarithms of numbers, we have reproduced, in Table III, the numbers alluded to by Döllen, as given by Wrangell, by means of which one easily passes from the logarithm of any number of seconds of arc to the corresponding logarithmic sine or tangent. If there be required the logarithmic sine or tangent of the angle x, which, expressed in seconds, we denote by x'', then we have

$$\text{Briggian log sine } x = \text{Briggian log } x'' + 4.685575 - \text{S}$$
$$\text{Briggian log tangent } x = \text{Briggian log } x'' + 4.685575 + \text{T}.$$

Our Table III gives the numbers S and T in units of the sixth place of decimals. It will be most convenient to omit the addition of the constant logarithm 4.685575, in cases where it is eventually eliminated from the result. In using six-figure logarithms the computer will find it advantageous to copy into Bremiker's six-figure tables the numbers S and T, given in the well-known edition of Vega by the same author.

The conversion of arc into time, and *vice versa*, should be avoided, if possible, by the use of tables in which the arguments are given, in parallel columns, both in time and arc.

GENERAL STAR CATALOGUE.

No.	A. B. J. C.	Name.	Mag.	α 1870.0	Annual variation.	δ 1870.0.	Annual variation.
				h. m. s.	s.	° ′ ″	″
1	A B J C	α Andromedæ	2	0 1 40	+ 3.1	+ 28 22 22	+ 19.9
2		β Cassiopeæ	2.4	0 2 13	+ 3.1	+ 58 26 0	+ 19.9
3	A B J C	γ Pegasi	2.6	0 6 33	+ 3.1	+ 14 27 39	+ 20.0
4	A B	β Hydræ	3	0 18 53	+ 3.3	— 77 59 14	+ 20.2
5	C	α Phœnicis	2	0 19 50	+ 3.0	— 43 0 38	+ 19.7
6	B	12 Ceti	6	0 23 24	+ 3.1	— 4 49 33	+ 19.9
7		δ Andromedæ	3.4	0 32 23	+ 3.2	+ 30 8 55	+ 19.7
8	A B J C	α Cassiopeæ	2.2–2.8	0 33 9	+ 3.4	+ 55 49 26	+ 19.8
9	A B J	β Ceti	2	0 37 4	+ 3.0	— 18 42 2	+ 19.8
10	A	21 Cassiopeæ	6	0 37 6	+ 3.8	+ 74 16 34	+ 19.7
11		γ Cassiopeæ	2	0 48 53	+ 3.5	+ 60 0 44	+ 19.6
12	A B C	ε Piscium	4	0 56 12	+ 3.1	+ 7 11 22	+ 19.5
13		η Ceti	3	1 2 3	+ 3.0	— 10 52 11	+ 19.2
14	C	β Andromedæ	2.4	1 2 28	+ 3.3	+ 34 55 53	+ 19.3
15	A B J C	α Ursæ Minoris	2	1 11 16	+ 20.1	+ 88 36 59	+ 19.1
16		δ Cassiopeæ	3	1 17 21	+ 3.8	+ 59 33 30	+ 18.9
17	A B	θ Ceti	3	1 17 31	+ 3.0	— 8 51 18	+ 18.7
18	A	A (38) Cassiopeæ	6	1 21 36	+ 4.4	+ 69 35 39	+ 18.7
19		γ Phœnicis	3	1 22 44	+ 2.6	— 43 59 2	+ 18.6
20	A B	η Piscium	3.6	1 24 32	+ 3.2	+ 14 40 30	+ 18.7
21	A B C	α Eridani	1	1 32 52	+ 2.2	— 57 53 51	+ 18.4
22	B	ν Piscium	4.6	1 34 40	+ 3.1	+ 4 49 43	+ 18.3
23	C	φ (54) Andromedæ	4.4	1 35 32	+ 3.7	+ 49 49 44	+ 18.4
24	A	ο Piscium	4	1 38 32	+ 3.2	+ 8 30 8	+ 18.3
25		ζ Ceti	3	1 45 3	+ 3.0	— 10 58 45	+ 17.9
26		ε Cassiopeæ	3.4	1 45 4	+ 4.2	+ 63 1 44	+ 18.0
27	A B C	β Arietis	2.6	1 47 28	+ 3.3	+ 20 10 17	+ 17.8
28	A	50 Cassiopeæ	4	1 52 23	+ 5.0	+ 71 47 24	+ 17.7
29		α Hydræ	3	1 54 40	+ 1.9	— 62 12 2	+ 17.6
30		γ Andromedæ	2.4	1 55 55	+ 3.6	+ 48 17 40	+ 17.6
31	A B J C	α Arietis	2	1 59 51	+ 3.4	+ 22 50 48	+ 17.2
32		β Trianguli	3	2 1 49	+ 3.5	+ 34 12 28	+ 17.3
33	A	ξ^1 (65) Ceti	4.4	2 6 7	+ 3.2	+ 8 14 8	+ 17.1
34	B	67 Ceti	6	2 10 30	+ 3.0	— 7 1 21	+ 16.7
35	C	ο Ceti	1.7–∞	2 12 47	+ 3.0	— 3 34 10	+ 16.6
36	A C	ι (35) Cassiopeæ	4	2 18 23	+ 4.8	+ 66 48 56	+ 16.5
37	B	ξ^2 Ceti	4	2 21 15	+ 3.2	+ 7 52 33	+ 16.3
38	A B J	γ Ceti	3.4	2 36 34	+ 3.1	+ 2 41 10	+ 15.4
39	C	41 Arietis	4	2 42 20	+ 3.5	+ 26 43 26	+ 15.2
40		η Eridani	3	2 50 4	+ 2.9	— 9 25 1	— 14.6
41		γ Persei	3	2 55 24	+ 4.3	+ 52 59 42	+ 14.6
42	A B J C	α Ceti	2.4	2 55 29	+ 3.1	+ 3 34 41	+ 14.4
43	C	β Persei	2.3...4.0	2 59 43	+ 3.9	+ 40 27 10	+ 14.3
44	A	48 Cephei	6	3 3 55	+ 7.3	+ 77 15 9	+ 13.9
45	B J	δ Arietis	4.4	3 4 12	+ 3.4	+ 19 14 0	+ 13.9
46	A	ζ Arietis	4.4	3 7 26	+ 3.4	+ 20 33 39	+ 13.7
47	A B J C	α Persei	2	3 15 3	+ 4.2	+ 49 23 45	+ 13.2
48		ε Eridani	3	3 26 45	+ 2.9	— 9 54 3	+ 12.4
49	A C	δ Persei	3	3 33 44	+ 4.2	+ 47 22 9	+ 11.9
50		δ Eridani	3	3 37 2	+ 2.9	— 10 12 20	+ 10.0
51	A B	η Tauri	3	3 39 46	+ 3.6	+ 23 42 3	+ 11.5
52	A	ζ Persei	3	3 45 58	+ 3.8	+ 31 29 42	+ 11.0
53		γ Hydræ	3	3 49 18	— 1.0	— 74 38 16	+ 10.9
54	A B	γ Eridani	3	3 51 58	+ 2.8	— 13 52 49	+ 10.5
55	B	o^1 Eridani	4.4	4 5 31	+ 2.9	— 7 10 42	+ 9.7
56	A C	γ Tauri	4	4 12 24	+ 3.4	+ 15 18 41	+ 9.1
57	A B	ε Tauri	3.6	4 21 3	+ 3.5	+ 18 53 22	+ 8.4
58	A B J C	α Tauri	1	4 28 28	+ 3.4	+ 16 14 45	+ 7.6
59		52 Eridani	3	4 2[illegible] 30	+ 2.3	— 30 49 47	+ 7.7
60		α Doradus	3	4 31 11	+ 1.3	— 55 18 53	+ 7.6
61	A	α (9) Camelopardalis	4	4 41 8	+ 5.9	+ 66 7 4	+ 6.8
62	C	π^1 Orionis	5	4 42 47	+ 3.3	+ 6 43 56	+ 6.7
63	A B	ι Aurigæ	3	4 48 32	+ 3.9	+ 32 57 27	+ 6.1
64	C	β (10) Camelopardalis	4	4 51 52	+ 5.3	+ 60 14 55	+ 6.0
65	A	11 Orionis	5	4 57 9	+ 3.4	+ 15 13 14	+ 5.4

GENERAL STAR CATALOGUE—Continued.

No.	A. B. J. C.	Name.	Mag.	α 1870.0.	Annual variation.	δ 1870.0.	Annual variation.
				h. m. s.	s.	° ′ ″	″
66	B	ε Leporis	3.6	4 59 57	+ 2.5	— 22 32 51	+ 5.1
67		β Eridani	3	5 1 28	+ 2.9	— 5 15 25	+ 5.0
68	A B J C	α Aurigæ	1	5 7 5	+ 4.4	+ 45 51 45	+ 4.2
69	A B J C	β Orionis	1	5 8 17	+ 2.9	— 8 21 15	+ 4.5
70	A B J C	β Tauri	2	5 18 5	+ 3.8	+ 28 29 41	+ 3.5
71	C	γ Orionis	2	5 18 10	+ 3.2	+ 6 13 49	+ 3.7
72	A	966 Groombridge	6.4	5 22 22	+ 8.0	+ 74 57 5	+ 3.3
73	A B C	δ Orionis	2.2...2.7	5 25 22	+ 3.1	— 0 23 53	+ 3.0
74	A B	α Leporis	3	5 27 0	+ 2.6	— 17 55 3	+ 2.9
75		ι Orionis	3	5 29 4	+ 2.9	— 5 59 50	+ 2.8
76	A B C	ε Orionis	2	5 29 37	+ 3.0	— 1 17 14	+ 2.6
77	C	ζ Orionis	2	5 34 12	+ 3.0	— 2 0 49	+ 2.3
78	A B C	α Columbæ	2	5 34 57	+ 2.2	— 34 8 40	+ 2.2
79		κ Orionis	2.6	5 41 35	+ 2.8	— 9 43 5	+ 1.7
80		β Columbæ	3	5 46 23	+ 2.1	— 35 49 14	+ 1.5
81	A B J C	α Orionis	1.0...1.4	5 48 8	+ 3.2	+ 7 22 49	+ 1.1
82		β Aurigæ	2	5 49 0	+ 4.4	+ 44 55 52	+ 1.0
83		θ Aurigæ	3	5 50 51	+ 4.1	+ 37 12 4	+ 0.8
84	B	ν Orionis	4.6	6 0 9	+ 3.4	+ 14 46 53	0.0
85	A	22 Camelopardalis	4.6	6 4 31	+ 6.6	+ 69 21 38	— 0.5
86	A B	μ Geminorum	3	6 15 6	+ 3.6	+ 22 34 38	— 1.4
87		ζ Canis Majoris	2.6	6 15 20	+ 2.3	— 30 0 28	— 1.3
88	C	β Canis Majoris	2.4	6 16 58	+ 2.6	— 17 53 36	— 1.4
89	A B C	α Argus	1	6 21 4	+ 1.3	— 52 37 32	— 1.8
90	A B	γ Geminorum	2.4	6 30 12	+ 3.5	+ 16 30 28	— 2.7
91		ν Argus	3	6 33 47	+ 1.8	— 43 4 57	— 2.9
92		Ξ Geminorum	3.4	6 35 59	+ 3.7	+ 25 15 26	— 3.0
93	A B	51 Cephei	5	6 38 42	+ 30.3	+ 87 14 23	— 3.4
94	A B J C	α Canis Majoris	1	6 39 26	+ 2.6	— 16 32 22	— 4.6
95	A B	ε Canis Majoris	1.6	6 53 31	+ 2.4	— 28 47 50	— 4.6
96	B	γ Canis Majoris	4.4	6 57 53	+ 2.7	— 15 26 35	— 5.0
97	A	δ Canis Majoris	2	7 3 6	+ 2.4	— 26 11 18	— 5.2
98	A B	δ Geminorum	3.4	7 12 21	+ 3.6	+ 22 13 8	— 6.2
99		π Argus	3.4	7 12 34	+ 2.1	— 36 51 55	— 6.2
100	A	Piazzi VII, 67	6	7 17 20	+ 6.3	+ 68 43 35	— 6.7
101		η Canis Majoris	2.6	7 18 57	+ 2.4	— 29 3 4	— 6.7
102	C	β Canis Minoris	3	7 20 4	+ 3.3	+ 8 32 5[illegible]	— 6.8
103	A B J C	α² Geminorum	1.6	7 26 18	+ 3.8	+ 32 10 15	— 7.4
104	A B J C	α Canis Minoris	1	7 32 30	+ 3.1	+ 5 33 22	— 8.9
105	A B J C	β Geminorum	1.4	7 37 22	+ 3.7	+ 28 20 16	— 8.3
106	C	ξ Argus	3.4	7 43 50	+ 2.5	— 24 32 7	— 8.7
107	A	φ Geminorum	5	7 45 32	+ 3.7	+ 27 5 58	— 8.9
108		χ Argus	3	7 53 28	+ 1.5	— 52 38 4	— 9.5
109	B	μ Cancri	5	7 55 32	+ 3.7	+ 28 9 23	— 9.8
110		ζ Argus	2.5	7 59 1	+ 2.1	— 39 38 20	— 10.0
111	A C	55 Camelopardalis 3 Ursæ Majoris	6	7 59 51	+ 6.1	+ 68 51 10	— 10.0
112	A B	ι (15) Argus	3	8 2 0	+ 2.6	— 23 55 52	— 10.1
113	C	γ² Argus	2	8 5 31	+ 1.8	— 46 57 19	— 10.5
114	C	β Cancri	3.6	8 9 27	+ 3.3	+ 9 35 3	— 10.7
115		ε Argus	2	8 19 51	+ 1.2	— 59 5 27	— 11.3
116	B	η Cancri	6	8 25 11	+ 3.5	+ 20 52 50	— 11.9
117	C	δ Hydræ	4.4	8 30 46	+ 3.2	+ 6 9 21	— 12.2
118	A B	ε Hydræ	3.4	8 39 53	+ 3.2	+ 6 53 39	— 12.9
119		δ Argus	3	8 41 7	— 1.7	— 54 14 0	— 13.1
120	A B J	ι Ursæ Majoris	3	8 50 18	+ 4.1	+ 48 33 0	— 13.8
121	A	σ² Ursæ Majoris	5	8 58 55	+ 5.4	+ 67 39 32	— 14.2
122	A	κ Cancri	5	9 0 42	+ 3.3	+ 11 11 22	— 14.2
123	C	ξ Cancri	5	9 1 53	+ 3.5	+ 22 34 13	— 14.2
124		λ Argus	3	9 3 13	+ 2.2	— 42 54 33	— 14.4
125	B	83 Cancri	6	9 11 43	+ 3.4	+ 18 15 17	— 15.1
126	C	β Argus	1	9 11 46	+ 0.8	— 69 10 57	— 14.8
127	A B C	ι Argus	2	9 13 37	+ 1.6	— 58 43 47	— 14.9
128		κ Argus	4	9 18 5	+ 1.9	— 54 27 26	— 15.3
129	A	1 Draconis	4.4	9 18 20	+ 9.2	+ 81 53 50	— 15.3
130	A B J C	α Hydræ	2.3...2.7	9 21 12	+ 3.0	— 8 5 47	— 15.4

GENERAL STAR CATALOGUE—Continued.

No.	A. B. J. C.	Name.	Mag.	α 1870.0.	Annual variation.	δ 1870.0.	Annual variation.
				h. m. s.	s.	° ′ ″	″
131	A	24 (d) Ursæ Majoris .	4.6	9 22 56	+ 5.4	+ 70 23 57	− 15.5
132	A B J	θ Ursæ Majoris . .	3	9 24 9	+ 4.1	+ 52 16 4	− 16.2
133	A B	ε Leonis	3	9 38 28	+ 3.4	+ 24 22 17	− 16.4
134		υ Argus	3	9 43 51	+ 1.5	− 64 28 10	− 16.6
135	A C	μ Leonis	4	9 45 22	+ 3.4	+ 26 37 4	− 16.7
136	B	π Leonis	5	9 53 21	+ 3.2	+ 8 40 0	− 17.1
137	A B J C	α Leonis . . .	1.4	10 1 27	+ 3.2	+ 12 36 6	− 17.4
138	A	32 Ursæ Majoris . .	6	10 8 34	+ 4.4	+ 65 45 19	− 17.8
139		ζ Leonis	3	10 9 28	+ 3.4	+ 24 3 52	− 17.7
140	A B J C	γ¹ Leonis	2	10 12 48	+ 3.3	+ 20 29 53	− 18.0
141		μ Ursæ Majoris . .	3	10 14 35	+ 3.6	+ 42 9 10	− 17.9
142	A	9 Draconis	4.6	10 23 58	+ 5.3	+ 76 22 52	− 18.4
143	A B C	ρ Leonis	4	10 25 58	+ 3.2	+ 9 58 28	− 18.4
144		θ Argus	3	10 38 19	+ 2.1	− 63 42 48	− 18.8
145	A B C	η Argus	1.....6	10 40 1	+ 2.3	− 59 0 3	− 18.8
146		μ Argus	3	10 41 12	+ 2.6	− 48 44 0	− 18.9
147	A B	l Leonis	5	10 42 25	+ 3.2	+ 11 13 56	− 18.9
148	C	ν Hydræ	3.4	10 43 3	+ 2.9	− 15 30 52	− 18.7
149	C	β Ursæ Majoris . .	2.4	10 53 59	+ 3.7	+ 57 4 43	− 19.2
150	A B J C	α Ursæ Majoris . .	2	10 55 41	+ 3.8	+ 62 27 7	− 19.4
151	B J	χ Leonis	5	10 58 19	+ 3.1	+ 8 2 17	− 19.4
152		ψ Ursæ Majoris . .	3	11 2 21	+ 3.4	+ 45 12 10	− 19.5
153	A B J	δ Leonis	2.4	11 7 12	+ 3.2	+ 21 14 8	− 19.6
154		θ Leonis	3.4	11 7 21	+ 3.1	+ 16 8 24	− 19.5
155	A B J	δ Crateris	3.4	11 12 51	+ 3.0	− 14 4 32	− 19.4
156	A	τ Leonis	5	11 21 15	+ 3.1	+ 3 34 19	− 19.8
157	A C	λ Draconis	3.4	11 23 39	+ 3.6	+ 70 2 52	− 19.9
158	A B	υ (91) Leonis . . .	4.6	11 30 18	+ 3.1	− 0 6 22	− 19.8
159	A B J C	β Leonis	2	11 42 26	+ 3.1	+ 15 17 56	− 20.1
160	J C	β Virginis	3.5	11 43 55	+ 3.1	+ 2 29 50	− 20.3
161	A B J C	γ Ursæ Majoris . .	2.4	11 46 59	+ 3.2	+ 54 25 3	− 20.0
162	A	ο Virginis	4	11 58 35	+ 3.1	+ 9 27 18	− 20.0
163	B	ε Corvi	3	12 3 26	+ 3.1	− 21 53 48	− 20.0
164	A	4 Draconis	4.6	12 6 5	+ 2.9	+ 78 20 18	− 20.1
165		δ Crucis	3	12 8 15	+ 3.1	− 58 1 29	− 20.0
166	C	δ Ursæ Majoris . .	3.4	12 8 59	+ 3.0	+ 57 45 16	− 20.1
167		γ Corvi	2	12 9 7	+ 3.1	− 16 49 10	− 20.0
168	A B	β Chamæleontis . .	5	12 10 46	+ 3.3	− 78 35 26	− 20.0
169	A B C	η Virginis	3.4	12 13 15	+ 3.1	+ 0 3 21	− 20.0
170	A B C	α¹ Crucis	1	12 19 23	+ 3.3	− 62 2 38	− 19.9
171	C	δ Corvi	2.4	12 23 9	+ 3.1	− 16 47 28	− 20.1
172		γ Crucis	2	12 23 58	+ 3.3	− 56 23 0	− 20.1
173	A B	β Corvi	2.4	12 27 34	+ 3.1	− 22 40 40	− 20.0
174	A	κ Draconis	3.4	12 27 55	+ 2.6	+ 70 30 17	− 19.9
175		γ Centauri	3	12 34 21	+ 3.3	− 48 14 45	− 19.9
176	B {	γ¹ Virginis	2.6	12 35 4	+ 3.0	− 0 44 12	− 19.8
177	J }	γ² Virginis	2.6	12 35 5	+ 3.0	− 0 44 6	− 19.8
178	C	β Crucis	2	12 40 8	+ 3.4	− 58 58 34	− 19.7
179	A	32 Camelopardalis .	4.6	12 48 12	+ 0.4	+ 84 7 9	− 19.6
180	C	ε Ursæ Majoris . .	2	12 48 18	+ 2.7	+ 56 39 56	− 19.7
181		δ Virginis	3	12 49 4	+ 3.0	+ 4 6 15	− 19.7
182	A B J	12 Canum Venat. . .	3	12 49 57	+ 2.8	+ 39 1 16	− 19.5
183	C	ε Virginis	2.6	12 55 43	+ 3.0	+ 11 39 32	− 19.5
184	A B	θ Virginis	4.4	13 3 13	+ 3.1	− 4 50 40	− 19.3
185		γ Hydræ	3	13 11 51	+ 3.2	− 22 29 2	− 19.1
186		ι Centauri	3	13 13 18	+ 3.4	− 36 1 34	− 19.1
187	A B J C	α Virginis	1	13 18 21	+ 3.2	− 10 28 55	− 18.9
188		ζ Ursæ Majoris . .	2	13 18 41	+ 2.4	+ 55 36 17	− 18.9
189	A B J C	ζ Virginis	3.4	13 28 4	+ 3.1	+ 0 4 11	− 18.5
190		ε Centauri	3	13 31 40	+ 3.7	− 52 48 15	− 18.6
191	A B J C	η Ursæ Majoris . .	2	13 42 25	+ 2.4	+ 49 57 47	− 18.1
192		ζ Centauri . . .	3	13 47 26	+ 3.7	− 46 38 50	− 18.0
193	A B J	η Bootis	3	13 48 30	+ 2.9	+ 19 3 1	− 18.2
194	A B C	β Centauri	1	13 54 40	+ 4.2	− 59 44 40	− 17.7
195	B	τ Virginis	4	13 55 2	+ 3.0	+ 2 10 28	− 17.6

GENERAL STAR CATALOGUE—Continued.

No.	A. B. J. C.	Name.	Mag.	α 1870.0.	Annual variation.	δ 1870.0.	Annual variation.
				h. m. s.	s.	° ′ ″	″
196	C	θ Centauri	3	13 59 3	+ 3.5	− 35 43 48	− 18.1
197	A C	α Draconis	3.4	14 0 52	+ 1.6	+ 64 59 50	− 17.4
198	A B J C	α Bootis	1	14 9 44	+ 2.7	+ 19 51 38	− 18.9
199	A	[illegible] Bootis	3.6	14 20 46	+ 2.0	+ 52 27 9	− 16.8
200	B	ρ Bootis	3.6	14 26 14	+ 2.6	+ 30 56 36	− 16.0
201		γ Bootis	2.6	14 26 51	+ 2.4	+ 38 52 40	− 16.0
202		η Centauri	3	14 27 15	+ 3.8	− 41 35 9	− 16.2
203	A	5 Ursæ Minoris	4.7	14 27 50	− 0.2	+ 76 16 25	− 16.0
204	A B C	α² Centauri	1	14 30 48	+ 4.0	− 60 17 39	− 15.0
205		α Lupi	3	14 33 16	+ 3.9	− 46 49 43	− 15.9
206	C	ξ Bootis	3.4	14 34 57	+ 2.9	+ 14 17 14	− 15.7
207	A B	ε Bootis	2.4	14 39 19	+ 2.6	+ 27 37 24	− 15.4
208	J	α¹ Libræ	6	14 43 27	+ 3.3	− 15 27 2	− 15.2
209	A B J C	α² Libræ	2.4	14 43 41	+ 3.3	− 15 29 59	− 15.2
210		β Lupi	3	14 50 2	+ 3.9	− 42 36 31	− 15.0
211		κ Centauri	3	14 50 43	+ 3.9	− 41 24 56	− 14.8
212	A B J C	β Ursæ Minoris	2	14 51 7	− 0.3	+ 74 41 11	− 14.8
213	A C	β Bootis	3	14 57 3	+ 2.3	+ 40 54 15	− 14.4
214	B J	ψ Bootis	4.4	14 58 53	+ 2.6	+ 27 27 22	− 14.2
215		γ Trianguli Australis	3	15 6 49	+ 5.5	− 68 11 47	− 13.9
216	A B	β Libræ	2	15 10 1	+ 3.2	− 8 54 5	− 13.6
217		δ Bootis	3	15 10 16	+ 2.4	+ 33 48 6	− 13.7
218	A	μ¹ Bootis	3.6	15 19 35	+ 2.3	+ 37 50 4	− 12.8
219	A C	[illegible] Ursæ Minoris	3	15 20 57	− 0.1	+ 72 17 48	− 12.8
220		ι Draconis	3	15 22 3	+ 1.3	+ 59 25 20	− 12.8
221		γ Lupi	3	15 26 29	+ 4.0	− 40 43 44	− 12.7
222		δ Serpentis	3.4	15 28 36	+ 2.9	+ 10 58 34	− 12.3
223	A B J C	α Coronæ Borealis	2	15 29 11	+ 2.5	+ 27 9 15	− 12.3
224	A B J C	α Serpentis	2.4	15 37 52	+ 3.0	+ 6 50 12	− 11.6
225		β Trianguli Australis	3	15 43 43	+ 5.2	− 63 1 30	− 11.7
226	A C	ε Serpentis	3.4	15 44 20	+ 3.0	+ 4 52 15	− 11.1
227	A B J	ζ Ursæ Minoris	4.4	15 48 45	− 2.3	+ 78 11 35	− 10.9
228		γ Serpentis	3.6	15 50 26	+ 2.8	+ 16 5 19	− 12.0
229		π Scorpii	3	15 51 0	+ 3.6	− 25 44 17	− 10.8
230	A	ε Coronæ Borealis	4	15 52 12	+ 2.5	+ 27 15 21	− 10.6
231	A	δ Scorpii	2.4	15 52 39	+ 3.5	− 22 14 57	− 10.6
232	A B	β¹ Scorpii	2	15 57 53	+ 3.5	− 19 26 50	− 10.2
233		θ Draconis	3.6	15 59 28	+ 1.1	+ 58 54 46	− 9.8
234	A	2320 Groombridge	5.6	16 5 58	+ 0.1	+ 68 9 10	− 9.5
235	A B	δ Ophiuchi	3	16 7 32	+ 3.1	− 3 21 27	− 9.6
236		ε Ophiuchi	3.4	16 11 26	+ 3.2	− 4 22 25	− 9.2
237	A	τ Herculis	3.4	16 15 50	+ 1.8	+ 46 37 27	− 8.8
238		γ Herculis	3	16 16 11	+ 2.6	+ 19 28 6	− 8.8
239	A B J C	α Scorpii	1.4	16 21 26	+ 3.7	− 26 8 26	− 8.4
240	A B	η Draconis	2.6	16 22 15	+ 0.8	+ 61 48 33	− 8.2
241		β Herculis	2.4	16 24 38	+ 2.6	+ 21 46 21	− 8.2
242	A	15 (A) Draconis	5	16 28 15	− 0.1	+ 69 2 58	− 7.8
243	A	ζ Ophiuchi	2.6	16 30 0	+ 3.3	− 10 18 5	− 7.6
244	A B C	α Trianguli Australis	2	16 34 56	+ 6.3	− 68 47 4	− 7.4
245	B J	ζ Herculis	2.6	16 36 23	+ 2.3	+ 31 50 24	− 6.8
246	A	η Herculis	3	16 38 26	+ 2.1	+ 39 10 16	− 7.0
247	C	ε Scorpii	3	16 41 45	+ 3.9	− 34 3 17	− 7.1
248		μ¹ Scorpii	3	16 42 4	+ 4.0	− 37 49 21	− 6.8
249		ζ² Scorpii	3	16 45 26	+ 4.2	− 42 8 14	− 6.9
250	A B J	κ Ophiuchi	3.4	16 51 31	+ 2.8	+ 9 34 45	− 5.9
251	C	ε Herculis	3.4	16 55 19	+ 2.3	+ 31 7 12	− 5.6
252	A	d Herculis	5	16 56 48	+ 2.2	+ 33 45 30	− 5.4
253	A B	ε Ursæ Minoris	4.4	16 59 23	− 6.4	+ 82 14 49	− 5.2
254		η Ophiuchi	2.4	17 2 53	+ 4.3	− 15 33 40	− 5.4
255		ζ Draconis	3	17 8 25	+ 0.2	+ 65 52 29	− 4.5
256	A B J C	α¹ Herculis	3.1…3.9	17 8 43	+ 2.7	+ 14 32 27	− 4.4
257		δ Herculis	3	17 9 41	+ 2.5	+ 24 59 38	− 4.6
258	B	θ Ophiuchi	3.4	17 14 2	+ 3.7	− 24 51 59	− 4.0
259		γ Aræ	3	17 14 28	+ 5.0	− 56 15 4	− 4.1
260		β Aræ	3	17 14 30	+ 5.0	− 55 24 13	− 4.2

GENERAL STAR CATALOGUE—Continued.

No.	A.	B.	J.	C.	Name.	Mag.	α 1870.0.	Annual variation.	δ 1870.0.	Annual variation.
							h. m. s.	s.	° ′ ″	″
261	A				b (44) Ophiuchi	5	17 18 26	+ 3.7	− 24 3 11	− 3.7
262					α Aræ	3	17 21 48	+ 4.6	− 49 46 10	− 3.6
263					λ Scorpii	3	17 24 47	+ 4.1	− 37 0 19	− 3.1
264	A	B	J		β Draconis	2.6	17 27 30	+ 1.4	+ 52 23 55	− 2.8
265					θ Scorpii	3	17 27 56	+ 4.3	− 42 54 47	− 3.1
266	A	B	J	C	α Ophiuchi	2	17 28 54	+ 2.8	+ 12 39 54	− 2.9
267					κ Scorpii	3	17 33 30	+ 4.1	− 38 57 38	− 2.5
268				C	β Ophiuchi	3	17 37 3	+ 3.0	+ 4 37 25	− 1.9
269	A				ω Draconis	5	17 37 43	− 0.4	+ 68 49 2	− 1.6
270	A	B	J		μ Herculis	3.4	17 41 22	+ 2.3	+ 27 47 54	− 2.4
271	A				ψ¹ Herculis, (pr.)	4.4	17 44 15	− 1.1	+ 72 12 43	− 1.6
272	A	B	J	C	γ¹ Draconis	2.4	17 53 35	+ 1.4	+ 51 30 18	− 0.6
273	A				γ² Sagittarii	3.4	17 57 27	+ 3.9	− 30 25 23	− 0.4
274	A	B			μ¹ Sagittarii	4	18 5 59	+ 3.6	− 21 5 25	+ 0.5
275	A	B			σ Octantis	6	18 6 19	+ 109.2	− 89 16 43	+ 0.6
276	A	B	J	C	δ Ursæ Minoris	4.4	18 14 16	− 19.4	+ 86 36 20	+ 1.3
277	A				η Serpentis	3	18 14 35	+ 3.1	− 2 55 49	+ 0.6
278					ε Sagittarii	2.6	18 14 52	+ 4.0	− 34 26 33	+ 1.2
279					λ Sagittarii	3	18 19 57	+ 3.7	− 25 29 30	+ 1.4
280	A				1 Aquilæ	4.4	18 28 8	+ 3.3	− 8 19 58	+ 2.2
281	A	B	J	C	α Lyræ	1	18 32 32	+ 2.0	+ 38 39 51	+ 3.1
282	A	B	J	C	β¹ Lyræ	3.5...4.5	18 45 17	+ 2.2	+ 33 12 47	+ 3.9
283	A				σ Sagittarii	2.4	18 47 12	+ 3.7	− 26 27 19	+ 4.0
284	A				50 Draconis	6	18 50 33	− 1.9	+ 75 16 44	+ 4.4
285					γ Lyræ	3.4	18 54 5	+ 2.2	+ 32 30 48	+ 4.6
286				C	ζ Sagittarii	3.4	18 54 20	+ 3.8	− 30 3 50	+ 4.6
287					λ Aquilæ	3.4	18 59 21	+ 3.2	− 5 4 30	+ 5.0
288	A	B			ζ Aquilæ	3	18 59 26	+ 2.8	+ 13 40 21	+ 5.1
289					π Sagittarii	3	19 2 2	+ 3.6	− 21 13 38	+ 5.3
290	A				d Sagittarii	5	19 10 2	+ 3.5	− 19 10 55	+ 6.1
291		B			ω Aquilæ	5.6	19 11 43	+ 2.8	+ 11 21 46	+ 6.2
292	A			C	δ Draconis	3	19 12 31	0.0	+ 67 25 58	+ 6.3
293	A				τ Draconis	5	19 18 2	− 1.1	+ 73 6 47	+ 6.8
294	A	B	J		δ Aquilæ.	3.4	19 18 57	+ 3.0	+ 2 51 28	+ 6.9
295				C	β Cygni	3	19 25 29	+ 2.4	+ 27 41 20	+ 7.3
296		B			h² Sagittarii	4.6	19 28 47	+ 3.7	− 25 10 3	+ 7.6
297	A				κ Aquilæ	5	19 29 54	+ 3.2	− 7 18 51	+ 7.7
298	A	B	J	C	γ Aquilæ	3	19 40 5	+ 2.9	+ 10 17 55	+ 8.5
299					δ Cygni	3	19 40 55	+ 1.9	+ 44 48 50	+ 8.5
300	A	B	J	C	α Aquilæ	1.4	19 44 26	+ 2.9	+ 8 31 37	+ 9.2
301					η Aquilæ	3.5...4.7	19 45 51	+ 3.1	+ 0 40 26	+ 8.8
302	A				ε Draconis	4	19 48 36	− 0.2	+ 69 56 11	+ 9.2
303	A	B	J	C	β Aquilæ	4	19 48 56	+ 2.9	+ 6 5 2	+ 8.7
304	A	B			λ Ursæ Minoris	6.4	19 54 17	− 59.2	+ 88 55 4	+ 9.6
305	A				τ Aquilæ	5.6	19 57 47	+ 2.9	+ 6 54 46	+ 9.9
306				C	θ Aquilæ	3	20 4 36	+ 3.1	− 1 12 17	+ 10.3
307			J		α¹ Capricorni	4.4	20 10 26	+ 3.3	− 12 54 28	+ 10.8
308	A	B	J	C	α² Capricorni	3.4	20 10 50	+ 3.3	− 12 56 44	+ 10.8
309	A				κ Cephei	4.4	20 13 13	− 1.9	+ 77 19 0	+ 11.0
310					β Capricorni	3	20 13 42	+ 3.4	− 15 11 22	+ 11.0
311	A	B		C	α Pavonis	2	20 15 21	+ 4.8	− 57 8 54	+ 11.1
312				C	γ Cygni	2.6	20 17 34	+ 2.2	+ 39 50 31	+ 11.3
313	A				π Capricorni	5	20 19 53	+ 3.4	− 18 38 8	+ 11.5
314		B			ρ Capricorni	5	20 21 26	+ 3.4	− 18 14 28	+ 11.6
315	A				ε Delphini	4	20 27 0	+ 2.9	+ 10 51 48	+ 12.0
316					α Indi	3	20 28 25	+ 4.3	− 47 44 31	+ 11.9
317	A				3241 Groombridge	6.4	20 30 33	− 0.2	+ 72 5 28	+ 12.2
318					β Pavonis	3	20 33 13	+ 5.5	− 66 39 58	+ 12.3
319				C	α Delphini	3.6	20 33 36	+ 2.8	+ 15 27 19	+ 12.4
320	A	B	J	C	α Cygni	1.6	20 37 0	+ 2.0	+ 44 49 0	+ 12.7
321					ε Cygni	2.6	20 40 57	+ 2.4	+ 33 29 5	+ 13.2
322	A				μ Aquarii	4.6	20 45 38	+ 3.2	− 9 28 9	+ 13.2
323		B			32 Vulpeculæ	5.4	20 49 1	+ 2.6	+ 27 33 52	+ 13.5
324	A				ν Cygni	4	20 52 20	+ 2.2	+ 40 40 4	+ 13.7
325	A				1879 Twelve year C.	6	20 53 24	− 2.5	+ 80 3 47	+ 13.7

GENERAL STAR CATALOGUE—Continued.

No.	A. B. J. C.	Name.	Mag.	α 1870.0.	Annual variation.	δ 1870.0.	Annual variation.
				h. m. s.	s.	° ′ ″	″
326	A B J	61[1] Cygni	5.4	21 1 4	+ 2.7	+ 38 6 41	+ 17.5
327	A B	ζ Cygni	3	21 7 24	+ 2.6	+ 29 41 42	+ 14.6
328	A B J C	α Cephei	2.6	21 15 28	+ 1.4	+ 62 2 6	+ 15.1
329		γ Pavonis	3	21 15 40	+ 5.1	− 65 57 11	+ 15.7
330	A	1 Pegasi	4.4	21 16 4	+ 2.8	+ 19 15 0	+ 15.2
331	A B C	β Aquarii	3	21 24 43	+ 3.2	− 6 8 30	+ 15.6
332	A B J	β Cephei	3	21 26 58	+ 0.8	+ 69 59 24	+ 15.7
333	A	ξ Aquarii	4.6	21 30 50	+ 3.2	− 8 26 9	+ 15.9
334	A B	ε Pegasi	2.4	21 37 48	+ 2.9	+ 9 16 49	+ 16.3
335	C	δ Capricorni	3	21 39 51	+ 3.3	− 16 42 54	+ 16.1
336	A	11 Cephei	5	21 40 0	+ 0.9	+ 70 42 46	+ 16.5
337		γ Gruis	3	21 46 3	+ 3.7	− 37 58 33	+ 16.5
338	A	μ Capricorni	5	21 46 12	+ 3.3	− 14 9 45	+ 16.8
339	B	16 Pegasi	5.4	21 47 9	+ 2.7	+ 25 18 52	+ 16.8
340	A	79 Draconis	6.4	21 51 15	+ 0.7	+ 73 5 14	+ 17.0
341	A B J C	α Aquarii	3	21 59 6	+ 3.1	− 0 57 1	+ 17.3
342	A B C	α Gruis	2	22 0 2	+ 3.8	− 47 35 20	+ 17.2
343	C	ζ Cephei	3.6	22 6 21	+ 2.1	+ 57 33 39	+ 17.6
344		α Tucanæ	3	22 9 34	+ 4.2	− 60 54 20	+ 17.7
345	A B	θ Aquarii	4.4	22 9 58	+ 3.2	− 8 25 46	+ 17.8
346	C	γ Aquarii	3.6	22 14 57	+ 3.1	− 2 2 28	+ 18.0
347	A	π Aquarii	4.6	22 18 38	+ 3.1	+ 0 43 6	+ 18.1
348	A B	η Aquarii	3.6	22 28 40	+ 3.1	− 0 47 12	+ 18.4
349	A	226 Cephei	5.4	22 29 59	+ 1.1	+ 75 33 23	+ 18.5
350		β Gruis	3	22 34 54	+ 3.6	− 47 33 40	+ 18.6
351	A B	ζ Pegasi	3.4	22 34 58	+ 3.0	+ 10 9 14	+ 18.7
352		η Pegasi	3	22 36 55	+ 2.8	+ 29 32 32	+ 18.7
353	A	ι Cephei	3.6	22 45 3	+ 2.1	+ 65 31 1	+ 18.8
354	A	λ Aquarii	4	22 45 50	+ 3.1	− 8 16 13	+ 19.1
355		δ Aquarii	3	22 47 45	+ 3.2	− 16 30 40	+ 19.1
356	A B J C	α Piscis Australis	1.4	22 50 28	+ 3.3	− 30 18 38	+ 19.0
357		β Pegasi	2.2..2.7	22 57 28	+ 2.9	+ 27 22 43	+ 19.5
358	A B J C	α Pegasi	2	22 58 17	+ 3.0	+ 14 30 24	+ 19.3
359	B J C	γ Piscium	4	23 10 26	+ 3.1	+ 2 34 20	+ 19.6
360	A	ο Cephei	5.6	23 13 18	+ 2.4	+ 67 24 0	+ 19.6
361	B	κ Piscium	4.6	23 20 16	+ 3.1	+ 0 32 39	+ 19.6
362	A	θ Piscium	4.4	23 21 22	+ 3.0	+ 5 39 54	+ 19.7
363	A B J	ι Piscium	4.4	23 33 16	+ 3.1	+ 4 55 18	+ 19.5
364	A B	γ Cephei	3.4	23 34 2	+ 2.4	+ 76 54 25	+ 20.1
365	B	δ Sculptoris	4.4	23 42 9	+ 3.1	− 28 50 55	+ 19.9
366	A	4163 Groombridge	7	23 48 32	+ 2.8	+ 73 41 12	+ 20.0
367	A B J	ω Piscium	4	23 52 38	+ 3.1	+ 6 8 36	+ 19.9
368	C	2 Ceti	4.4	23 57 5	+ 3.1	− 18 3 34	+ 20.1

www.ingramcontent.com/pod-product-compliance
Lightning Source LLC
Chambersburg PA
CBHW022040080426
42733CB00007B/915